W0106133

# SPRINGER TRACTS IN MODERN PHYSICS

Ergebnisse
der exakten Natur-
wissenschaften

Volume  56

Editor: G. Höhler

Editorial Board: P. Falk-Vairant  S. Flügge  J. Hamilton
F. Hund  H. Lehmann  E. A. Niekisch  W. Paul

Springer-Verlag Berlin Heidelberg GmbH 1971

*Manuscripts for publication should be addressed to:*

G. HÖHLER, Institut für Theoretische Kernphysik der Universität, 75 Karlsruhe, Kaiserstraße 12

*Proofs and all correspondence concerning papers in the process of publication should be addressed to:*

E. A. NIEKISCH, Kernforschungsanlage Jülich, Institut für Technische Physik, 517 Jülich, Postfach 365

ISBN 978-3-662-15594-3      ISBN 978-3-540-36399-6 (eBook)
DOI 10.1007/978-3-540-36399-6

This work is subject to copyright. All rights are reserved, whether the whole or part of the material is concerned, specifically those of translation, reprinting, re-use of illustrations, broadcasting, reproduction by photocopying machine or similar means, and storage in data banks.

Under § 54 of the German Copyright Law where copies are made for other than private use, a fee is payable to the publisher, the amount of the fee to be determined by agreement with the publisher. © by Springer-Verlag Berlin Heidelberg 1971. Originally published by Springer-Verlag Berlin Heidelberg New York in 1971. Softcover reprint of the hardcover 1st edition 1971. Library of Congress Catalog Card Number 25-9130.

The use of general descriptive names, trade names, trade marks, etc. in this publication, even if the former are not especially identified, is not be taken as a sign that such names, as understood by the Trade Marks and Merchandise Marks Act, may accordingly be used freely by anyone.

# The Method of the Correlation Function in Superconductivity Theory

G. Lüders and K.-D. Usadel

## Contents

Contents                                                                      3

## Frequently Used Symbols

This list does not contain symbols which appear in a single section only. Single letters are sometimes used in a different meaning than the one given below.

| | |
|---|---|
| $a$ | thickness of a metallic film or layer |
| $B_j$ | boundary of a metal labeled $j$, in a sample consisting of several metals |
| $F_1\left(\dfrac{\xi_0}{l}\right), F_2\left(\dfrac{\xi_0}{l}\right)$ | Eqs. (17.41) and (17.49), Figs. 3 and 4 |
| $G_\omega^{(0)}(r, r')$ | Green's function of normal electrons: Section 1 |
| $g$ | BCS coupling constant |
| $g_\omega(r, v; r')$ | distribution function: Section 5, Eq. (10.5) |
| $H_\omega(r), \boldsymbol{H}_\omega(r)$ | Section 14a |
| $i(r)$ | electric current density: Eq. (1.32), Section 18 |
| $K_\omega(r, r')$ | kernel in the gap equation (2.15), (10.1): Eqs. (2.16), (4.39), (10.4) |
| $\mathbb{K}_\omega$ | operator which corresponds to $K_\omega(r, r')$:Eq. (11.17) |
| $k$ | wave number vector or classical momentum vector, in particular of conduction electrons |
| $k$ | parameter in Eq. (19.3) |
| $l$ | electronic mean free path: Eq. (12.15) |
| $l_{\text{tr}}$ | transport length: Eq. (13.27) |
| $m$ | electronic mass |
| $m^*$ | effective mass: Section 4d |
| $m_{jk}, \left(\dfrac{1}{m}\right)_{jk}$ | mass tensor, reciprocal mass tensor: Section 4d |
| $N, N(r), N_j$ | density of (orbital) electronic states: Section 3b, Eq. (10.3) |
| $\mathbb{N}$ | operator which corresponds to $N(r)$: Eq. (11.34) |
| $n, n(r)$ | impurity concentration (number per volume) |
| $n$ | surface normal unit vector (pointing to the exterior), interface normal unit vector |
| $p^{\cdot}$ | diffuseness of the reflexion of electrons by a surface: Eqs. (6.11), (10.24), (15.42) |
| $p'$ | further parameter characterizing the reflexion properties of a surface: Eq. (15.54) |
| $R(r; v_{\text{out}}, v_{\text{in}})$ | reflexion kernel: Section 6, Eqs. (10.16), (10.26) |
| $T, T_c$ | temperature, transition temperature of a single metal or of metals in contact, in the absence of a magnetic field: Sections 11c, 23b |
| $T_j(r; v_{j\text{out}}, v_{k\,\text{in}})$ | transmission kernel: Section 6b, Eq. (10.26) |
| $t_j, t_j'$ | parameters characterizing the transmission properties of an interface: Eqs. (15.73), (15.77) |
| $V_j$ | volume (interior) of a metal labeled $j$, in a sample consisting of several metals |
| $v$ | velocity, modulus of $\boldsymbol{v}$, usually equal to Fermi velocity |
| $\boldsymbol{v}$ | velocity vector, modulus usually equal to Fermi velocity |
| $v(k)$ | Eq. (4.40) |
| $v_{\text{in}}, v_{\text{out}}$ | velocity vector of electron approaching a surface or interface (incoming electrons) or of electrons leaving a surface or interface (outgoing electrons) |
| $\gamma$ | Euler's constant: Eq. (11.43) |
| $\Delta(r)$ | pair potential: Section 1a |
| $\Delta_\omega(r, v)$ | distribution function: Eq. (11.22) |
| $\hat{\delta}(a, b)$ | delta function of the direction of vectors: Eq. (4.14) |

| | |
|---|---|
| $\varepsilon_{jkl}$ | totally antisymmetric Ricci tensor, $\varepsilon_{123} = +1$ |
| $\zeta(n)$ | Riemann's zeta function: Eqs. (13.30), (13.12), (13.54) |
| $\zeta_0(T)$ | Eq. (17.20) |
| $\zeta_\omega$ | Eq. (12.14) |
| $\eta(k)$ | momentum dependent energy, in particular of conduction electrons; $\eta(k) = 0$: Fermi surface |
| $\theta$ | scattering angle |
| $\kappa_\omega$ | eigenvalue of $\mathbb{K}_\omega$: Eq. (17.8) |
| $\lambda$ | Eq. (17.16) |
| $\lambda(T)$ | Eq. (13.1); Eqs. (13.13) and (13.26) (Ginzburg-Landau approximation); Eq. (14.31), Fig. 2 (diffusion approximation) |
| $\mu$ | chemical potential, or parameter in Eq. (19.3), or Eq. (14.33), Fig. 2 |
| $\xi, \varepsilon, \eta$ | dimensionless variables: Eqs. (21.13), (21.22) |
| $\xi(T)$ | Eq. (13.2) |
| $\xi_0(T)$ | Eq. (13.11) |
| $\xi_\omega$ | Eq. (12.4) |
| $\sigma$ | scattering cross section; $\sigma_{tr}$: Eq. (13.22); $\sigma^{(1)}$: Eq. (13.20); $\sigma^{(2)}$: Eq. (13.40); $\sigma_1, \sigma_{sp}$: Section 9 b |
| $\dfrac{d\sigma(\theta)}{d\Omega}, \dfrac{d\sigma(v, v')}{d\Omega}$ | differential scattering cross section |
| $\Phi_0$ | unit of quantized flux: after Eq. (19.9) (symbol not always used with this meaning) |
| $\chi\left(\dfrac{\xi_0}{l}\right)$ | Eq. (13.28), Fig. 1 |
| $\psi(n)$ | digamma function: Eq. (13.33) |
| $\omega$ | odd multiple of $\pi T$: Section 1 b |
| $\omega_D$ | Debye frequency, cut-off frequency |
| $dr$ | vectorial line element |
| $d^2 r$ | vectorial surface or interface element, $\parallel n$ |
| $d^3 r$ | scalar volume element ($> 0$) |
| $dS$ | surface element of the energy surface, $\eta(k) = $ const, in $k$ space, in particular of the Fermi surface |
| $d\Omega$ | element of solid angle, in particular with respect to the direction of $v$ |
| $\tilde{\partial}$ | gauge invariant gradient: Eqs. (5.16), (10.6) |

# Introduction

In 1964 *de Gennes* [1] initiated a method for the theoretical analysis of second kind transitions between the normal and the superconducting state of metallic samples. A correlation function played a central part in this method. *De Gennes'* method was subsequently further developed and applied to various situations by him and by others.

In the present review paper a report shall be given on published and unpublished work done in this direction at Göttingen during the last few years. It is hoped that such a systematic presentation might be helpful to other workers in this field. Though correlation functions play a minor part in the formulation to be presented in this report, we want to retain the name "method of the correlation function". This baptization should be regarded as an hommage to *de Gennes*.

The review paper is divided into three chapters. In Chapter I, the general formulation is first developed on a somewhat intuitive basis and afterwards as far as possible proved from the microscopic theory. This chapter is mainly based upon papers by the senior author [2–6] which have been published in German. Chapter II is concerned with a general discussion of the basic formulation and with the derivation [7] of approximative formulations (mainly Ginzburg-Landau approximation [8] and diffusion approximation [1, 9, 10]). These approximative schemes are useful for the analysis of special problems. In Chapter III finally, the formulations developed in Chapter II are applied to the calculation of critical magnetic fields for various geometries and to a sample consisting of superposed layers of different metals.

The method of the correlation function may be characterized as a semi-microscopic method. It is based upon the microscopic BCS picture in its more flexible Gorkov version which, therefore, is surveyed in Section 1 of Chapter I. This scheme is specialized to second kind transitions in Section 2. The quantity of central importance in Gorkov's theory is the pair potential, $\Delta(r)$. It vanishes in the normal state. It goes continuously to zero if a second kind transition is approached from the superconducting side. Gorkov's self-consistency condition for the pair potential ("gap equation") becomes a linear integral equation for the pair potential, at a second kind transition. The main goal of the method of the correlation function is the calculation of the kernel, $K_\omega(r, r')$, entering into this integral equation. General properties of the kernel are summarized in Section 3.

Following *de Gennes* [1], the kernel is expressed by a correlation function for electrons in Section 4. Arguments are put forward that, not the quantum mechanical, but the classical correlation function has to be used in this context. The classical correlation function may be derived from a distribution function in phase space. A Boltzmann equation for the distribution function is suggested in Section 5. Whereas a strictly classical treatment may be given only if the system is invariant under time reversal (no magnetic field and no paramagnetic impurities present), a simple and straightforward extension to the case of a magnetic field is also suggested in Section 5. The Boltzmann equation should be valid inside a homogeneous metal. It has to be supplemented by boundary conditions for the distribution function on external surfaces and, if the sample consists of several metals, on interfaces. The boundary conditions follow immediately from the classical picture. They are formulated in Section 6.

The Boltzmann equation which was first suggested on intuitive grounds is subsequently proved in Sections 7 (free electrons) and 8 (electrons in a periodic lattice) on the basis of the microscopic Gorkov theory. Also arguments in favour of the special semi-microscopic boundary condition for specular reflexion will be given. The analysis is extended to the case of paramagnetic impurities (where the subsystem of conduction electrons violates time reversal invariance) in Section 9.

The reader not interested in the somewhat intricate proofs on the basis of the microscopic theory is advised to skip Sections 7–9 and to go immediately over to Chapter II. It is also possible to start reading with Chapter II since the semi-microscopic formulation (consisting of Boltzmann equation and boundary conditions), which was put forward and derived in Chapter I, is summarized in Section 10 of Chapter II.

The early work reported in Chapter I was stimulated by discussions with *G. Eilenberger, J. Hajdu, W. Moormann,* and *E. Schöler.*

The Göttingen work is restricted to second kind transitions. An interesting extension to the general case with $\Delta(r)$ not being infinitesimally small has been given by *Eilenberger* [11]. It seems, however, an open question how surfaces and interfaces may be incorporated into this more general scheme.

*We should like to summarize what we regard the main advantages of the method of the correlation function:*

*1. A formulation using concepts of classical mechanics, like particle orbits and distribution functions, appeals to physical imagination.*

*2. The microscopic theory, in the presence of impurities, requires a summation of contributions from infinitely many graphs. This summation is done once for ever in the proof of the Boltzmann equation.*

*3. Surfaces and interfaces with stochastic elements seem not amenable to a microscopic quantum-mechanical treatment, but stochastic elements are easily incorporated into the semi-microscopic boundary conditions to be used in the method of the correlation function.*

As already said above, Chapter II starts in Section 10 with a summary of the results obtained in the first chapter. Some general aspects of the theory are discussed in Section 11. This section is partly based upon work by *H. J. Sommers*. Classical particle orbits are exhibited in various ways in Section 12.

The general formulation is still rather clumsy if specific problems shall be solved. Approximative schemes known from the literature are, therefore, derived in the next sections. The Ginzburg-Landau approximation [8] is valid for temperatures close to the transition temperature (in the absence of a magnetic field). The linearized Ginzburg-Landau equation, a second order linear differential equation for the pair potential, $\Delta(r)$, is derived in Section 13. Boundary conditions which supplement this equation are obtained [12] in Section 15. The diffusion approximation [9, 10] is, on the other hand, valid for metals with sufficiently short electronic mean free path. The diffusion equation, a second order differential equation for the kernel, $K_\omega(r, r')$, is derived in Section 14. Boundary conditions are obtained [13] in Section 16. A special method [14] for samples consisting of a single metal with isotropic impurity scattering is exposed in Section 18. This method to some extent unites Ginzburg-Landau and diffusion approximation.

General expressions for the electric current density, which goes continuously to zero together with the pair potential in the transition, are derived in Section 18.

*The derivation of boundary conditions shows that the customary requirements for the validity of the Ginzburg-Landau approximation (temperature close to transition temperature) or the diffusion approximation (electronic mean free path sufficiently short) are not sufficient in finite samples. Also the linear dimensions of the samples, including radii of curvature of surfaces and interfaces, have to be sufficiently big. A more quantitative version of this condition is given in the respective sections.*

Chapter III starts in Section 19 with a discussion of the bulk critical magnetic field, $H_{c2}$. Critical magnetic fields of samples consisting of a single metal in special geometries are studied in the next three sections: the surface normal critical field, $H_{c\perp}$, of semi-infinite samples with plane boundary and of layers with two plane parallel boundaries in Section 20; the surface parallel critical field, $H_{c3}$, of semi-infinite samples with plane boundary in Section 21; the surface parallel critical field, $H_{c\parallel}$, of plane layers in Section 22. The methods of Chapter II are mainly used to derive some corrections to the results obtained with the help of the Ginzburg-Landau and diffusion approximation.

Samples consisting of several metals in electric contact are studied in Section 23. Actually the analysis will be restricted to two plane super-posed layers with sufficiently short electronic mean free path so that the diffusion approximation is valid. Only the simplest case (so-called Cooper limit) is studied regarding its transition temperature and its critical magnetic fields $H_{c\perp}$ and $H_{c\parallel}$.

Chapter III is to some extent based upon mostly unpublished work by *K. D. Harms, A. Helms, P. Richter, H. Schultens, H. J. Sommers,* and *J. Steppeler.*

*All theoretical work done so far on samples consisting of a single metal suggests the conclusion that every stochastic element in the orbital motion of the electrons leads to an enhancement of critical magnetic fields for second kind transitions. These stochastic elements may be located in the interior (impurities which scatter the electrons) and on the surface (diffuseness of the scattering process). We quote the main examples for the general statement:*

*1. The bulk critical magnetic field, $H_{c2}$, for second kind transitions grows with increasing concentration of impurities (Section 19).*

*2. Critical magnetic fields of finite samples (like $H_{c\perp}$, $H_{c3}$, and $H_{c\parallel}$) become bigger with increasing diffuseness of the electron scattering by surfaces (Sections 20–22). This influence is the less pronounced the shorter the electronic mean free path in the metal is. It does not appear in the Ginzburg-Landau and diffusion approximations.*

*No convincing general proof for the enhancement of critical fields by stochastic elements has so far been found.*

Some work reported in this review paper has also been done by others, in particular by *de Gennes* [10, 15] himself. But contributions by other authors [16–21] should also be mentioned. In these papers some of the Göttingen work on the influence of diffusely scattering surfaces on critical magnetic fields was actually anticipated.

We wish to thank Mrs. *I. Lüders* and Dipl.-Phys. *H. J. Sommers* for their assistance in correcting the proofs.

# I. General Theory

## 1. Survey of Microscopic Superconductivity Theory

### a) Cooperative Pairing

The peculiar behaviour of superconducting metals is believed to be a consequence of the state of its conduction electrons. In the normal state of a metal, the electrons, at least approximately, form a free Fermi gas. But electron pairs appear whenever the metal becomes superconducting (BCS theory [22]). The pairs are Bosons. They condensate in macroscopic number into a single one-pair state. But the pair picture should not be strained since the spatial extension of a pair (as given by one of the correlation lengthes [23] occurring in superconductivity theory) is big as compared to the average distance between two pairs.

Pairing requires an attractive interaction between electrons which is due to a polarization of the lattice or, in quantum-mechanical language, to an exchange of phonons between the two members of a pair [22]. This phonon-mediated attraction is almost pointlike in space and strongly retarded in time since the electron velocity at the Fermi surface (the Fermi velocity) is big as compared to the phonon velocity: a virtual phonon is dropped by an electron at some position and does not move much until it is picked up some time later by another electron. The repulsive Coulomb interaction between the electrons is shielded and, therefore, also practically pointlike. The net interaction is attractive in metals which may become superconducting at sufficiently low temperatures.

Pairing alone does not suffice. The pairs have to assemble in a single one-pair state. This condensation, which is a cooperative phenomenon, may be understood as the result of the action of a kind of mean field. This mean field was introduced by *Gorkov* [24] into his flexible version of the microscopic theory of superconductivity. Various names have been given to it: pair potential, gap function, order parameter. The pair potential is produced by the electrons already paired. On the other hand, it forces the pairs into a single one-pair state. There would be no cooperative pair potential without the primary attractive electron-electron interaction which is due to the exchange of phonons.

In order to turn to a more quantitative formulation of the theory we first introduce the function

$$G_{\alpha\alpha'}(\boldsymbol{r}, \boldsymbol{r}') := \langle \psi_{\alpha'}^*(\boldsymbol{r}') \, \psi_\alpha(\boldsymbol{r}) \rangle , \tag{1.1}$$

where the angular brackets on the righthand side denote an average with respect to the grand canonical ensemble. The symbols in the bracket are, in second quantization language, operators for the creation of an electron with spin $\alpha'$ at the position $r'$ and for the annihilation of an electron with spin $\alpha$ at the position $r$. All one-electron properties may be calculated from the function $G_{\alpha\alpha'}(r, r')$.

In the superconducting state, there exists another non-vanishing expectation value of two electron operators, i.e.

$$F_{\alpha\alpha'}(r, r') := \varepsilon_{\alpha\beta} \langle \psi_{\alpha'}^*(r') \psi_{\beta}^*(r) \rangle. \tag{1.2}$$

This anomalous expectation value, which usually originates from an approximative treatment of the basic equations, gets its exact meaning through the concept of quasi-average [25]. The symbol $\varepsilon_{\alpha\beta}$ denotes the real skew-symmetric $2 \times 2$ matrix*

$$\varepsilon_{\alpha\beta} := \begin{pmatrix} 0 & 1 \\ -1 & 0 \end{pmatrix}. \tag{1.3}$$

It is

$$\varepsilon_{\alpha\beta}\varepsilon_{\beta\gamma} = -\delta_{\alpha\gamma}, \tag{1.4}$$

where $\delta_{\alpha\gamma}$ is the Kronecker symbol. Here and in Eq. (1.2), a summation is extended over spin subscripts appearing twice. The complex conjugate of the angular bracket in Eq. (1.2) has its direct physical meaning as the wave functions of the two partners of each pair. At the point of coincidence $(r' = r)$ this is the one-pair wave function of the cooperatively condensed pairs.

The pair potential, $\Delta(r)$, is given in terms of this wave function by

$$\Delta(r) \delta_{\alpha\beta} := g F_{\alpha\beta}^*(r, r), \tag{1.5}$$

where $g$ denotes the coupling constant of the primary electron-electron interaction. $g > 0$ means attraction. It seems to be an open question whether there exist intrinsic normal conductors with repulsive interaction ($g < 0$). In a sample consisting of several metals in electric contact, the coupling constant becomes position dependent: $g$ has to be replaced by $g(r)$. The asterisk on the righthand side of Eq. (1.5) denotes complex conjugation.

The introduction of a pair potential depending upon a single position variable (instead of two) requires the model of a strictly pointlike electron-electron interaction. Only such an interaction may be characterized by a single coupling constant, $g$. A modification of Eq. (1.5) is still required because of the retardation of the phonon-mediated electron-electron interaction. It shall be introduced in Eq. (1.7). That the righthand side of Eq. (1.5) is really proportional to the Kronecker symbol, $\delta_{\alpha\beta}$, is a consequence of the exclusion principle.

---

\* $F_{\alpha\alpha'}$ is usually defined without an $\varepsilon_{\alpha\beta}$ on the righthand side of Eq. (1.2). Our definition has some formal advantages.

## b) Gorkov's Equations

It follows from the fact, that electrons are Fermions, that the expectation values $G_{\alpha\alpha'}(r, r')$ and $F_{\alpha\alpha'}(r, r')$ may be written as

$$G_{\alpha\alpha'}(r, r') = \lim_{\eta \to +0} T \sum_{\omega} e^{i\omega\eta} G_{\omega\alpha\alpha'}(r, r'),$$

$$F_{\alpha\alpha'}(r, r') = T \sum_{\omega} F_{\omega\alpha\alpha'}(r, r'),$$

(1.6)

where the summation with respect to the frequency (or energy) variable, $\omega$, is extended from $-\infty$ to $+\infty$ over all odd integral multiples of $\pi T$ ($T$ = temperature with Boltzmann's constant put equal to one). The retardation of the electron-electron interaction is now somewhat artificially introduced into Eq. (1.5) by a cut-off,

$$\Delta(r)\, \delta_{\alpha\beta} = g\, T \sum_{\omega = -\omega_D}^{+\omega_D} F^*_{\omega\alpha\beta}(r, r).$$

(1.7)

$\omega_D$ is a frequency of the order of the Debye frequency. This procedure, which neglects all details of the phonon spectrum and of the electron-phonon interaction, is hoped to lead to reasonable results for so-called weak-coupling superconductors but it is not expected to be good for strong-coupling superconductors. The cut-off will usually be indicated by a dash on the summation symbol so that Eq. (1.7) reads

$$\Delta(r)\, \delta_{\alpha\beta} = g\, T \sum_{\omega}{}' F^*_{\omega\alpha\beta}(r, r).$$

(1.8)

The somewhat intricate problem of a position dependent cut-off for several metals in electric contact shall be postponed to Section 10.

In a suitable mean field approximation, the normal Green's function, $G_{\omega\alpha\alpha'}(r, r')$, and the anomalous Green's function, $F_{\omega\alpha\alpha'}(r, r')$, obey Gorkov's equations [24] (with $\hbar = 1$)

$$\left.\begin{aligned}
(i\omega + \mu)\, G_{\omega\alpha\alpha'}(r, r') - \int d^3 r'' \langle r\alpha | h | \alpha'' r'' \rangle\, G_{\omega\alpha''\alpha'}(r'', r') \\
+ \Delta(r)\, F_{\omega\alpha\alpha'}(r, r') = \delta_{\alpha\alpha'}\delta(r - r'), \\
(i\omega - \mu)\, F_{\omega\alpha\alpha'}(r, r') + \int d^3 r'' \langle r\alpha | \tilde{h} | \alpha'' r'' \rangle\, F_{\omega\alpha''\alpha'}(r'', r') \\
+ \Delta^*(r)\, G_{\omega\alpha\alpha'}(r, r') = 0.
\end{aligned}\right\}$$

(1.9)

In the normal state of the conduction electrons,

$$F^{(0)}_{\omega\alpha\alpha'}(r, r') \equiv 0,\ \Delta(r) \equiv 0,$$

(1.10)

they reduce to

$$(i\omega + \mu)\, G^{(0)}_{\omega\alpha\alpha'}(r, r') - \int d^3 r'' \langle r\alpha | h | \alpha'' r'' \rangle\, G^{(0)}_{\omega\alpha''\alpha'}(r'', r') = \delta_{\alpha\alpha'}\delta(r - r')$$

(1.11)

The quantity $\mu$ denotes the chemical potential (Fermi energy). $\langle r\alpha|h|\alpha''r''\rangle$ is a matrix element of the Hermitian one-electron Hamiltonian. Hartree-Fock terms due to interactions are neglected.

$$\langle r\alpha|\tilde{h}|\alpha''r''\rangle := -\varepsilon_{\alpha\beta}\langle r\beta|h|\beta''r''\rangle^*\varepsilon_{\beta''\alpha''} \tag{1.12}$$

is a matrix element of the time reversed one-electron Hamiltonian.

The action of the pair potential upon the electrons is exhibited in Eqs. (1.9). It is, in particular, seen from the second equation that non-vanishing pair potential leads to

$$F_{\omega\alpha\alpha'}(r, r') \not\equiv 0 \tag{1.13}$$

and, therefore, to a superconducting state with cooperatively paired electrons. Eq. (1.8) states how the pair potential is produced by the electron pairs. This equation, which is Gorkov's self-consistency condition, supplements Eqs. (1.9). Eq. (1.8) is also called gap equation.

For given one-electron Hamiltonian and given pair potential, the Green's functions are (for $\omega \neq 0$) uniquely determined by Eqs. (1.9). They satisfy the symmetry relations

$$\begin{aligned} G_{\omega\alpha'\alpha}(r', r) &= G^*_{-\omega\alpha\alpha'}(r, r'), \\ F_{\omega\alpha'\alpha}(r', r) &= -\varepsilon_{\alpha\gamma}F_{-\omega\gamma\gamma'}(r, r')\,\varepsilon_{\gamma'\alpha'}\,. \end{aligned} \tag{1.14}$$

For the proof of the uniqueness theorem we first notice that the difference of two solutions to Eqs. (1.9) obeys the same equations with vanishing righthand sides. Dropping the second spatial and spinorial arguments, we denote solutions to the homogeneous Gorkov's equations by $G_{\omega\alpha}(r)$ and $F_{\omega\alpha}(r)$, respectively. From a combination of these equations and the complex conjugate ones it is easily seen that it is

$$2i\omega \int \sum_{\alpha} (|G_{\omega\alpha}(r)|^2 + |F_{\omega\alpha}(r)|^2)\,\mathrm{d}^3r = 0, \tag{1.15}$$

which proves the assertion. A similar procedure of multiplying Gorkov's equations by appropriate Green's functions, adding, subtracting, and integrating leads to Eq. (1.14). The inhomogeneous part of Eqs. (1.9) is decisive for the proof.

In many cases of practical interest, the one-electron Hamiltonian is assumed to be spin independent,

$$\langle r\alpha|h|\alpha''r''\rangle =: \langle r|h|r''\rangle\,\delta_{\alpha\alpha''}\,. \tag{1.16}$$

In the presence of a magnetic field this can be true only if the action of the magnetic field upon the intrinsic magnetic moments is neglected, as is usually done in superconductivity theory. If Eq. (1.16) holds, also the time reversed Hamiltonian,

$$\langle r\alpha|\tilde{h}|\alpha''r''\rangle =: \langle r|\tilde{h}|r''\rangle\,\delta_{\alpha\alpha''}\,, \tag{1.17}$$

is spin independent. It is

$$\langle r|\tilde{h}|r''\rangle = \langle r|h|r''\rangle^* = \langle r''|h|r\rangle\,. \tag{1.18}$$

Under the assumption (1.16), the Green's functions become spin independent as well,

$$G_{\omega\alpha\alpha'}(r, r') = :G_\omega(r, r')\,\delta_{\alpha\alpha'},\ F_{\omega\alpha\alpha'}(r, r') = :F_\omega(r, r')\,\delta_{\alpha\alpha'}\,. \qquad (1.19)$$

They are the unique solutions to the spin independent Gorkov's equations [24]*,

$$(i\omega + \mu)\,G_\omega(r, r') - \int d^3 r'' \langle r|h|\,r'' \rangle\,G_\omega(r'', r') + \Delta(r)\,F_\omega(r, r') = \delta(r - r'),$$
$$(i\omega - \mu)\,F_\omega(r, r') + \int d^3 r'' \langle r|\tilde{h}|\,r'' \rangle\,F_\omega(r'', r') + \Delta^*(r)\,G_\omega(r, r') = 0\,. \qquad (1.20)$$

The gap equation (1.8) is replaced by

$$\Delta(r) = g\,T\,\sum_\omega{}'\,F_\omega^*(r, r)\,. \qquad (1.21)$$

The symmetry relations (1.14) become

$$G_\omega(r', r) = G^*_{-\omega}(r, r'), \qquad F_\omega(r', r) = F_{-\omega}(r, r')\,. \qquad (1.22)$$

The spin independent Green's function of normal electrons obeys

$$(i\omega + \mu)\,G_\omega^{(0)}(r, r') - \int d^3 r'' \langle r|h|\,r'' \rangle\,G_\omega^{(0)}(r'', r') = \delta(r - r')\,. \qquad (1.23)$$

The formulations are somewhat abstract as long as the physical model (or: the one-electron Hamiltonian) has not been specified. A particularly simple and instructive model is that of free electrons of mass $m$ and charge $-e$ in a magnetic field, $H(r)$, with vector potential $A(r)$,

$$H(r) = \operatorname{curl} A(r)\,. \qquad (1.24)$$

The interaction between the magnetic field and the spins of the electrons shall be neglected. In this case it is (with $c = 1$)

$$\left.\begin{aligned} \langle r|h|\,r'' \rangle &= -\frac{1}{2m}\left(\frac{\partial}{\partial r} + ieA(r)\right)^2 \delta(r - r''),\\[2mm] \langle r|\tilde{h}|\,r'' \rangle &= -\frac{1}{2m}\left(\frac{\partial}{\partial r} - ieA(r)\right)^2 \delta(r - r'')\,. \end{aligned}\right\} \qquad (1.25)$$

Introducing these matrix elements into Eq. (1.20) we obtain the free electron Gorkov's equations

$$\left[i\omega + \mu + \frac{1}{2m}\left(\frac{\partial}{\partial r} + ieA(r)\right)^2\right]G_\omega(r, r') + \Delta(r)\,F_\omega(r, r') = \delta(r - r'),$$
$$\left[i\omega - \mu - \frac{1}{2m}\left(\frac{\partial}{\partial r} - ieA(r)\right)^2\right]F_\omega(r, r') + \Delta^*(r)\,G_\omega(r, r') = 0\,. \qquad (1.26)$$

---

* *Gorkov* [24] uses the notation $F_\omega^+(r, r')$ for our $F_\omega(r, r')$.

It is seen that, under a gauge transformation

$$A(r) \to A'(r) = A(r) + \frac{\partial U(r)}{\partial r}, \tag{1.27}$$

the Green's functions transform in the following manner

$$\begin{aligned} G_\omega(r, r') \to G'_\omega(r, r') &= G_\omega(r, r') \exp[-ie(U(r) - U(r'))], \\ F_\omega(r, r') \to F'_\omega(r, r') &= F_\omega(r, r') \exp[ie(U(r) + U(r'))]. \end{aligned} \tag{1.28}$$

The pair potential is transformed as

$$\Delta(r) \to \Delta'(r) = \Delta(r) \exp[-ie(U(r) - U(r'))]. \tag{1.29}$$

The magnetic field acting upon the electrons is a sum,

$$H(r) = H_{\text{ext}}(r) + H_{\text{int}}(r), \tag{1.30}$$

of an external part, $H_{\text{ext}}(r)$, produced by coils or other equipment, and of an internal part, $H_{\text{int}}(r)$, generated by persistent electric currents in the superconductor. It is

$$\text{curl}\, H_{\text{int}}(r) = 4\pi i(r) \tag{1.31}$$

with the electric current density in the superconductor given by

$$\begin{aligned} i(r) &= \frac{ie}{2m} \left\langle \psi_\alpha^*(r)\left(\frac{\partial}{\partial r} + ieA(r)\right)\psi_\alpha(r) - \text{h.c.} \right\rangle \\ &= 2\frac{ie}{2m}\left(\frac{\partial}{\partial r} - \frac{\partial}{\partial r'} + 2ieA(r)\right) T\sum_\omega G_\omega(r, r')|_{r'=r}. \end{aligned} \tag{1.32}$$

The factor of two behind the second equality sign counts the number of spin states. The current due to the magnetic moments of the electrons is consistently neglected in Eq. (1.32). For a stationary current density we expect

$$\text{div}\, i(r) = 0 \tag{1.33}$$

to hold. This relation is a consequence of Eqs. (1.32), (1.26), (1.22), and (1.21). Because of the cut-off in Eq. (1.21) it is, however, required that $G_\omega(r, r')$ be replaced by $G_\omega^{(0)}(r, r')$ outside the interval $-\omega_D < \omega < +\omega_D$. The current due to $G_\omega^{(0)}(r, r')$ vanishes if the diamagnetic current in the Landau sense is neglected. This is usually done in superconductivity theory. Then, also the sum in Eq. (1.32) has to be cut off at $\pm\omega_D$.

That electric current density and internal magnetic field may be treated as classical quantities is not trivial. The physical reason for this assumption is that the current is produced by electron pairs which have condensated in macroscopic number into a single one-pair state. There occur now two self-consistency problems in the procedure of solving Gorkov's equations. One is connected with the pair potential [Eq. (1.21)] and the other with the magnetic field [Eqs. (1.31) and (1.32)].

It is not difficult to include a one-particle potential, $V(r)$, in the Hamiltonian,

$$\langle r|h| r''\rangle = \left[-\frac{1}{2m}\left(\frac{\partial}{\partial r} + ieA(r)\right)^2 + V(r)\right]\delta(r - r''),$$
$$\langle r|\tilde{h}| r''\rangle = \left[-\frac{1}{2m}\left(\frac{\partial}{\partial r} - ieA(r)\right)^2 + V(r)\right]\delta(r - r'').$$
(1.34)

Gorkov's equations (1.26) are modified in an obvious manner whereas Eqs. (1.28), (1.29), (1.32), and (1.33) remain valid.

We conclude this survey with a few more technical remarks. The unique solution to the inhomogeneous equation

$$(i\omega + \mu) X_\omega(r) - \int d^3r' \langle r|h| r'\rangle X_\omega(r') = A_\omega(r)$$
(1.35)

is given by

$$X_\omega(r) = \int d^3r' G_\omega^{(0)}(r, r') A_\omega(r')$$
(1.36)

with $G_\omega^{(0)}(r, r')$ defined by Eq. (1.23). Similar but more complicated expressions hold for spin dependent solutions to equations containing a spin dependent Hamiltonian. Eqs. (1.35) and (1.36) may be used to replace Gorkov's equations (1.20) by a set of integral equations,

$$G_\omega(r, r') = G_\omega^{(0)}(r, r') - \int G_\omega^{(0)}(r, r'') \Delta(r'') F_\omega(r'', r') d^3r'',$$
$$F_\omega(r, r') = \int G_\omega^{(0)*}(r, r'') \Delta^*(r'') G_\omega(r'', r') d^3r''.$$
(1.37)

These equations may be solved in an iterative manner with the result

$G_\omega(r, r')$
$$= G_\omega^{(0)}(r,r') - \int G_\omega^{(0)}(r,r'')\Delta(r'')G_\omega^{(0)*}(r'',r''')\Delta^*(r''')G_\omega^{(0)}(r''',r')d^3r''d^3r''' + \cdots,$$
$F_\omega(r, r')$
$$= \int G_\omega^{(0)*}(r,r'')\Delta^*(r'')G_\omega^{(0)}(r'',r')d^3r'' + \cdots.$$
(1.38)

If the electric current density is calculated according to Eq. (1.32), the first term on the righthand side of the first Eq. (1.38) leads to a diamagnetic contribution (which shall be neglected) whereas the current due to superconductivity is contained in the other terms.

## 2. Thermodynamics. Second Kind Transitions

*a) Phenomenological Thermodynamics*

Before continuing the microscopic theory we shall look upon super-conductivity from the point of view of phenomenological thermodynamics though statistical thermodynamics already entered into the microscopic theory through Eqs. (1.1) and (1.2). The macroscopic state of a simply connected reversible superconducting sample with no current fed into it is completely determined by external parameters, the temperature, $T$, and the magnetic field, $H_{ext}(r)$. The methods of phenomenological thermodynamics may, therefore, be applied. Normal state and super-conducting state may be regarded as different modifications or thermodynamic phases of the metal. In multiply connected samples, the physical state depends not only upon the external parameters but also upon the magnetic fluxes trapped by the various holes. The application of statistical and phenomenological thermodynamics to such samples meets perhaps with some doubt.

If temperature and magnetic field are used as the external thermodynamic parameters, the corresponding thermodynamic potential is called the Gibbs free energy. The superconducting state is stable whenever its Gibbs free energy is smaller than that of the normal state. A phase transition between normal and superconducting state may occur whenever the Gibbs free energies of the two states are equal. Therefore, the difference between the Gibbs free energies of the superconducting and the normal state is a quantity of particular interest. It shall be called $\mathscr{G} = \mathscr{G}[T, H_{ext}(r)]$. The normal state is stable for $\mathscr{G} > 0$. The superconducting state is stable for $\mathscr{G} < 0$. A phase transition occurs for $\mathscr{G} = 0$.

It follows from phenomenological thermodynamics that it is

$$\delta \mathscr{G} = -\mathscr{S} \delta T - \frac{1}{4\pi} \int H_{int}(r) \cdot \delta H_{ext}(r) \, d^3 r, \qquad (2.1)$$

where $\mathscr{S}$ is the entropy difference and $H_{int}(r)$ has been explained in connection with Eqs. (1.30) and (1.31). If $H_{int}(r)$ would, in the normal state, not be put equal to zero (Landau diamagnetism not neglected), $H_{int}(r)$ in Eq. (2.1) would have to be interpreted as the difference between these fields in the superconducting and the normal state. The integration in the magnetic term is extended, not only over the sample, but over the whole space.

If a transition occurs in a homogeneous magnetic field, the geometry and temperature dependent strength of this field is called a critical magnetic field. Using $H_{cx}(T)$ as a common notation for such critical fields we may write

$$H_{ext}(r) = H_{cx}(T) \, \hat{H}_{ext} \qquad (2.2)$$

where $\hat{H}_{ext}$ is the unit vector in the direction of the external magnetic field. The jump of the entropy at the transition (latent heat divided by $T$) is given by

$$\mathcal{S} = -\frac{1}{4\pi}\, \frac{\mathrm{d}H_{cx}(T)}{\mathrm{d}T}\, \hat{H}_{ext} \cdot \int H_{int}(r)\,\mathrm{d}^3 r\,. \qquad (2.3)$$

Eq. (2.3) follows immediately from Eq. (2.1) since it is $\delta\mathcal{G} = 0$ along the transition curve.

A phase transition is called of the first kind if it is connected with a latent heat. It is called of the second kind if no latent heat occurs. In the first case it is $\mathcal{S} \neq 0$ at the transition whereas the entropy difference, $\mathcal{S}$, vanishes in the second case. If $H_{int}(r)$ vanishes not only in the normal state but also in the superconducting state right at the transition, it is seen from Eq. (2.3) that $\mathcal{S}$ vanishes and that the transition, therefore, is of the second kind. If $H_{int}(r)$ jumps discontinuously, $\mathcal{S}$ will usually be different from zero so that the transition is of the first kind.

We give a few examples for samples consisting of a single metal. The transition without external magnetic field at the transition temperature, $T_c$, is of the second kind since no magnetic field is involved. A typical transition of the first kind occurs at the thermodynamic critical field, $H_c(T)$, in a superconductor of the first kind. The transition is, however, of the second kind at the field $H_{c2}(T)$ in a superconductor of the second kind since the magnetic field does not jump in this case. It should be noted that we regard as a superconducting state in this case what usually is called mixed state. Whether a particular metal is a superconductor of the first or of the second kind depends upon whether $H_c(T)$ is bigger or smaller than $H_{c2}(T)$. The considerations do not apply to the critical field $H_{c1}(T)$ since no transition between normal and superconducting state occurs at this field.

### b) Statistical Thermodynamics

A connection between the thermodynamic considerations of Section 2a and the microscopic theory presented in Section 1 is obtained by introducing the quantity

$$\tilde{\mathcal{G}} := \tilde{\mathcal{G}}_0 + \frac{1}{8\pi} \int H_{int}^2(r)\,\mathrm{d}^3 r \qquad (2.4)$$

with $\tilde{\mathcal{G}}_0$ given by

$$\tilde{\mathcal{G}}_0 := \int \left\{ \frac{|\varDelta(r)|^2}{g(r)} - 2\varDelta(r) \int_0^1 \mathrm{d}\lambda F(\lambda; r, r) \right\} \mathrm{d}^3 r\,. \qquad (2.5)$$

The integration in the magnetic field term is again extended over the whole space whereas the integration in $\tilde{\mathscr{G}}_0$ is extended over the sample only. The function $F(\lambda; r, r')$ is defined as a sum [cf. Eq. (1.6)] of functions $F_\omega(\lambda; r, r')$ which are solutions to the Gorkov equations (1.20) with $\Delta(r)$ replaced by $\lambda\Delta(r)$. The sum has again to be cut off at $\omega = \pm\omega_D$. The problem of a position dependent cut-off is postponed to Section 10.

In Eqs. (2.4) and (2.5), not only the external parameters, $T$ and $H_{ext}(r)$, are assumed to be given. $\tilde{\mathscr{G}}$ is also a real functional of the pair potential, $\Delta(r)$, and the internal magnetic field, $H_{int}(r)$, or, rather, of the vector potential, $A(r)$ [cf. Eqs. (1.30) and (1.24)]. Both $\Delta(r)$ and $A(r)$ appear in Eq. (1.20) and lead to a unique determination of $F_\omega(\lambda; r, r')$ and so of $F(\lambda; r, r)$. In particular it is

$$F(\lambda; r, r) \equiv 0 \quad \text{for} \quad \Delta(r) \equiv 0. \tag{2.6}$$

We now assert that pair potential and vector potential have, for $T$ and $H_{ext}(r)$ given, to be chosen in such a way that $\tilde{\mathscr{G}}$ is stationary with respect to variations of $\Delta(r)$,

$$\frac{\delta\tilde{\mathscr{G}}}{\delta\Delta(r)} = 0 = \frac{\delta\tilde{\mathscr{G}}}{\delta\Delta^*(r)}, \tag{2.7}$$

and of $A(r)$,

$$\frac{\delta\tilde{\mathscr{G}}}{\delta A(r)} = 0. \tag{2.8}$$

For the equilibrium situation defined in this way, $\tilde{\mathscr{G}}$ becomes equal to the Gibbs free energy difference, $\mathscr{G}$, introduced in connection with Eq. (2.1).

No complete proof of the assertions shall here be given. It can indeed be shown that Eq. (2.7) is equivalent to Eq. (1.21) and that Eq. (2.8) is equivalent to Eq. (1.31). The two supplementary conditions to be used in connection with the Gorkov equations are, therefore, contained in the requirement that $\tilde{\mathscr{G}}$ be stationary with respect to variations of $\Delta(r)$ and $A(r)$. A comparison of Eqs. (2.8) and (1.31) leads to the expression

$$i(r) = -\frac{\delta\tilde{\mathscr{G}}_0}{\delta A(r)} \tag{2.9}$$

for the electric current density in the superconducting metal. It might again be mentioned that the Landau diamagnetic current has been neglected.

Let $\Delta(r)$ and $A(r)$ now be chosen in such a way that the stationarity conditions (2.7) and (2.8) are satisfied. $\tilde{\mathscr{G}}$ and $\mathscr{G}$ become identical. If the external parameters, $T$ and $H_{ext}(r)$, are then varied also $\Delta(r)$ and $A(r)$ will have to be varied but, because of Eqs. (2.7) and (2.8), only the variations of the external parameters contribute to

$$\delta \mathscr{G} = \delta \tilde{\mathscr{G}} = \frac{\partial \tilde{\mathscr{G}}_0}{\partial T} \delta T + \int \frac{\delta \tilde{\mathscr{G}}_0}{\delta A(r)} \cdot \delta A(r) \, \mathrm{d}^3 r \,. \tag{2.10}$$

Here it is

$$\delta H_{\text{ext}}(r) = \text{curl} \, \delta A(r) \,. \tag{2.11}$$

Comparison of Eq. (2.10) with Eq. (2.1) leads to

$$\mathscr{S} = -\frac{\partial \tilde{\mathscr{G}}_0}{\partial T} \,. \tag{2.12}$$

The differentiation has to be done with respect to the explicit $T$ dependence only which enters through the variable $\omega$ [cf. comments to Eq. (1.6)]. There is also a contribution from the cut-off at $\pm \omega_D$, which shall not be discussed. The second term on the righthand side of Eq. (2.10) is easily identified with the second term on the righthand side of Eq. (2.1), if use is made of Eqs. (2.9) and (1.31).

A second kind transition between normal and superconducting state may now also be characterized on the microscopic level. Such a transition occurs whenever $\Delta(r)$, which vanishes in the normal state, becomes infinitesimally small in the superconducting state right at the transition,

$$\Delta(r) \rightarrow 0 \,. \tag{2.13}$$

Indeed, the electric current density vanishes in this case as a consequence of Eqs. (1.32) and (1.38) (diamagnetic current neglected!). Then also $H_{\text{int}}(r)$ vanishes in the superconducting state right at the transition. The transition is of the second kind as shown above in connection with Eq. (2.3). The reverse statement, that $\Delta(r)$ vanishes continuously at a second kind transition, cannot easily be proved. It shall, nevertheless, assumed to be true. Finally it should be mentioned that $\mathscr{S} = 0$ (condition for second kind transition) for $\Delta(r) \rightarrow 0$ also follows directly from Eq. (2.12).

The formulation of the BCS-Gorkov theory of superconductivity in terms of a variational principle has been suggested to us primarily by *G. Eilenberger*.

## c) Second Kind Transitions

From now on "second kind transition" and "$\Delta(r) \rightarrow 0$ at the transition" shall be used as synonyms. In this case, the second Eq. (1.38) is simplified,

$$F_\omega(r, r') = \int G_\omega^{(0)*}(r, r'') \, \Delta^*(r'') \, G_\omega^{(0)}(r'', r') \, \mathrm{d}^3 r'' \,. \tag{2.14}$$

Higher order terms in $\Delta(r)$ are strictly negligeable. The gap equation (1.21) may be written as

$$\Delta(r) = g \, T \int \sum_\omega' K_\omega(r, r') \, \Delta(r') \, \mathrm{d}^3 r' \,. \tag{2.15}$$

The kernel, $K_\omega(r, r')$, of this linear homogeneous integral equation is given by

$$K_\omega(r, r') := G_\omega^{(0)}(r, r') \, G_\omega^{(0)*}(r', r). \qquad (2.16)$$

It is determined by the quantum mechanics of electrons in their normal state. Much of the work to be reported on in this review is concerned with the calculation of this kernel.

A transition of the second kind may occur whenever Eq. (2.15) admits a non-trivial solution for appropriately chosen values of the external parameters, $T$ and $H_{ext}(r)$. But not every such solution is of physical relevance. So, for given temperature and geometry, only the highest magnetic field, for wich a non-trivial solution to Eq. (2.15) exists, is a critical field for a second kind transition.

The electric current density in the metal may be calculated from Eqs. (1.32) and (1.38). If the diamagnetic current is again neglected it is

$$i(r) = -\frac{ie}{m} \left( \frac{\partial}{\partial r} - \frac{\partial}{\partial r'} + 2ie\,A(r) \right)$$

$$T \sum_\omega{}' \int G_\omega^{(0)}(r, r'') \, \Delta(r'') \, G_\omega^{(0)*}(r'', r''') \, \Delta^*(r''') \, G_\omega^{(0)}(r''', r') \, \mathrm{d}^3 r'' \mathrm{d}^3 r''' \big|_{r'=r}. \qquad (2.17)$$

Higher order terms in $\Delta(r)$ and $\Delta^*(r)$ may consistently be discarded. The current obviously vanishes in the transition $[\Delta(r) \to 0]$. The current pattern is, nevertheless, of interest since it permits conclusions about the magnetic field pattern close to the transition.

The connection with the thermodynamic treatment presented above is obtained if the quantity $\tilde{\mathcal{G}}$ as defined by Eqs. (2.4) and (2.5) is evaluated in the limit $\Delta(r) \to 0$. First we have to write down the quantity $F_\omega(\lambda; r, r)$ which according to Eqs. (2.14) and (2.16) is given by

$$F_\omega(\lambda; r, r) = \lambda \int K_\omega(r, r') \, \Delta(r') \, \mathrm{d}^3 r'. \qquad (2.18)$$

Inserting this expression into Eqs. (2.4) and (2.5) we obtain

$$\tilde{\mathcal{G}} = \Phi + \cdots. \qquad (2.19)$$

with

$$\Phi = \int \frac{|\Delta(r)|^2}{g} \, \mathrm{d}^3 r - \int \Delta^*(r) \, T \sum_\omega{}' K_\omega(r, r') \, \Delta(r') \, \mathrm{d}^3 r \, \mathrm{d}^3 r'. \qquad (2.20)$$

The dots in Eq. (2.19) stand for terms of higher order in $\Delta(r)$ and for magnetic terms which are also of higher order. $\Phi$ is a real bilinear functional of the pair potential, $\Delta(r)$. The reality of $\Phi$ follows from the Hermiticity,

$$K_\omega(r', r) = K_\omega^*(r, r'), \qquad (2.21)$$

of the kernel, $K_\omega(r, r')$, which is an immediate consequence of its definition (2.16).

It is now not difficult to prove the stationarity condition (2.7), or

$$\frac{\delta\Phi}{\delta\Delta(r)} = 0 = \frac{\delta\Phi}{\delta\Delta^*(r)}. \tag{2.22}$$

Evaluation of this equation indeed leads to the gap equation (2.15). We add the remark that it is

$$\Phi = 0 \tag{2.23}$$

at the transition as a consequence of Eq. (2.15) even if $\Delta(r)$ is not chosen as an infinitesimal quantity. Eq. (2.9) reduces to

$$i(r) = -\frac{\delta\Phi}{\delta A(r)} = \int \Delta^*(r')\, T \sum_\omega{}' \frac{\delta K_\omega(r', r'')}{\delta A(r)} \Delta(r'')\, d^3r'\, d^3r''. \tag{2.24}$$

This relation will be confirmed below in small print.

The analysis shall be restricted to the model of free electrons though the inclusion of a one-particle potential would not lead to any difficulties. The normal conducting Green's function, $G_\omega^{(0)}(r', r'')$, is the unique solution to the equation

$$\left[ i\omega + \mu + \frac{1}{2m}\left( \frac{\partial}{\partial r'} + ieA(r') \right)^2 \right] G_\omega^{(0)}(r', r'') = \delta(r' - r''), \tag{2.25}$$

which follows from Eqs. (1.23) and (1.25). Variation with respect to $A(r)$ leads to

$$\left[ i\omega + \mu + \frac{1}{2m}\left( \frac{\partial}{\partial r'} + ieA(r') \right)^2 \right] \frac{\delta G_\omega^{(0)}(r', r'')}{\delta A(r)}$$
$$= -\frac{ie}{2m}\left[ \delta(r' - r)\left( \frac{\partial}{\partial r'} + ieA(r') \right) + \left( \frac{\partial}{\partial r'} + ieA(r') \right)\delta(r' - r) \right] G_\omega^{(0)}(r', r''). \tag{2.26}$$

Here, use was made of the relation

$$\frac{\delta A_j(r')}{\delta A_k(r)} = \delta_{jk}\delta(r' - r), \tag{2.27}$$

where $j$ and $k$ represent Cartesian subscripts. Eq. (2.26) is readily solved with the help of Eqs. (1.35) and (1.36) with the result

$$\frac{\delta G_\omega^{(0)}(r', r'')}{\delta A(r)} = -\frac{ie}{2m}\left[ G_\omega^{(0)}(r', r)\left( \frac{\partial}{\partial r} + ieA(r) \right) G_\omega^{(0)}(r, r'') \right.$$
$$\left. - G_\omega^{(0)}(r, r'')\left( \frac{\partial}{\partial r} - ieA(r) \right) G_\omega^{(0)}(r', r) \right]. \tag{2.28}$$

Insertion of this expression into Eq. (2.24) with

$$\frac{\delta K_\omega(r', r'')}{\delta A(r)} = \frac{\delta G_\omega^{(0)}(r', r'')}{\delta A(r)}\, G_\omega^{(0)*}(r'', r') + G_\omega^{(0)}(r', r'')\, \frac{\delta G_\omega^{(0)*}(r'', r')}{\delta A(r)} \tag{2.29}$$

immediately leads to the current density as given in Eq. (2.17).

The formal treatment of second kind transitions is evidently rather simple in principle. There are, however, also physical problems involved which we shall now shortly discuss. When the pair potential becomes small or even infinitesimally small in magnitude, its influence upon the condensation of pairs into a single one-pair state becomes very weak. Fluctuations, which were entirely neglected in the mean field (Gorkov) approach, might become important. The problem of phase transitions of the second kind and of fluctuations is still under vivid discussion in the physics community. It does not seem clear at present how reliable the results are which are obtained by a mean field approach. This is in particular so for results concerning finite samples.

On the other hand, the mean field approach leads to certain transition temperatures also for finite samples. The experimentalist measures transition temperatures always on finite samples. We propose to identify the transition temperatures and critical fields calculated on the basis of the Gorkov theory with those measured in the experiment. But it should be kept in mind that the logical basis for this identification is somewhat obscure.

## 3. General Properties of the Kernel. Density of States

### a) General Properties of $K_\omega(r, r')$

In this section we want to summarize and to derive some general properties of the kernel $K_\omega(r, r')$, which has been introduced in Section 2 [Eq. (2.16)] in connection with second kind transitions between normal and superconducting state

It follows immediately from the definition that $K_\omega(r, r')$ is Hermitian; cf. Eq. (2.21). It further follows from the first Eq. (1.22), which is also valid for $G_\omega^{(0)}(r, r')$, that it is

$$K_\omega(r, r') = K_{-\omega}(r, r'), \tag{3.1}$$

so that $K_\omega(r, r')$ depends only upon the modulus of $\omega$. In the presence of a magnetic field, the kernel transforms in the following way,

$$K_\omega(r, r') \to K'_\omega(r, r') = K_\omega(r, r') \exp[2ie(U(r') - U(r))], \tag{3.2}$$

under gauge transformations, Eq. (1.27). This result follows from Eq. (1.28). The gap equation (2.15) is gauge invariant if Eq. (1.29) is taken into account.

With no magnetic field present [and $A(r) \equiv 0$] and without paramagnetic impurity scattering, the one-electron Hamiltonian is invariant under time reversal,

$$\langle r|h| r'' \rangle = \langle r|h| r'' \rangle^*. \tag{3.3}$$

It then follows that it is

$$G_\omega^{(0)*}(r, r') = G_{-\omega}^{(0)}(r, r')$$     (3.4)

since $G_\omega^{(0)}(r, r')$ is uniquely determined by Eq. (1.23). Combining Eq. (3.4) with the first Eq. (1.22), we obtain

$$G_\omega^{(0)}(r', r) = G_\omega^{(0)}(r, r')$$     (3.5)

so that

$$K_\omega(r, r') = |G_\omega^{(0)}(r, r')|^2 = K_\omega(r', r)$$     (3.6)

is a real, symmetric, and non-negative quantity.

In order to derive a result [1] which is less evident we introduce the complete and orthonormal set of one-electron wave functions, $w_m(r)$, defined by the Schrödinger equation

$$\int \langle r|h| \, r'\rangle \, w_m(r') \, d^3 r' = (\eta_m + \mu) \, w_m(r) \,.$$     (3.7)

The quantity $\eta_m$ evidently is the difference between the one-electron energy and the chemical potential (Fermi energy), $\mu$. It follows from the orthonormality and completeness of the wave functions that it is

$$\sum_m w_m(r) \, w_m^*(r') = \delta(r - r')$$     (3.8)

where the subscript, $m$, is, for the sake of simplicity, regarded as a discrete variable. If Eq. (3.8) is used in conjunction with Eqs. (3.7) and (1.23), it follows that the Green's function may be written as

$$G_\omega^{(0)}(r, r') = \sum_m \frac{w_m(r) \, w_m^*(r')}{i\omega - \eta_m} \,.$$     (3.9)

Eq. (2.16) leads to [1]

$$K_\omega(r, r') = \sum_{mn} \frac{w_m(r) \, w_m^*(r') \, w_n^*(r') \, w_n(r)}{(\eta_m - i\omega) \, (\eta_n + i\omega)} \,.$$     (3.10)

If invariance under time reversal holds in the sense of Eq. (3.3), also the functions $w_m^*(r)$ satisfy Eq. (3.7). Also they form a complete and orthonormal set so that Eq. (3.10) may now be written in the form

$$K_\omega(r, r') = \sum_{mn} \frac{w_m(r) \, w_m^*(r') \, w_n(r') \, w_n^*(r)}{(\eta_m - i\omega) \, (\eta_n + i\omega)} \,.$$     (3.11)

Integration with respect to one of the coordinates, say $r$, leads to

$$\int K_\omega(r, r') \, d^3 r = \sum_n \frac{|w_n(r')|^2}{\eta_n^2 + \omega^2} \,.$$     (3.12)

The summation on the righthand side may approximately be replaced by an integration with respect to the variable $\eta$, if all linear dimensions of the sample are big as compared to the Fermi wavelength (lattice constant). If $|\omega|$ is sufficiently small, the slowly varying numerator may be evaluated at the Fermi energy ($\eta = 0$) and the $\eta$ integration may be extended from $-\infty$ to $+\infty$. In this way, the sum rule [1]

$$\int K_\omega(r, r')\, d^3 r = \frac{\pi N(r')}{|\omega|}$$  (3.13)

is obtained. The symbol

$$N(r') = \mathcal{N}\, \overline{|w_n(r')|^2}$$  (3.14)

denotes the so-called density of states of normal conducting electrons at the position $r'$ which shall be discussed in some detail further below. In Eq. (3.14), $\mathcal{N}$ is the number of orbital electron states per energy interval near the Fermi energy. The bar on the righthand side denotes an average on the Fermi surface.

b) Density of States

The density of states is, in the model of free electrons, given by

$$N = \frac{m^2 v}{2\pi^2}.$$  (3.15)

Here, $m$ is the mass of the electron and $v$ the Fermi velocity. In the presence of a periodic lattice potential, which is produced by the ions, it is

$$N = \frac{1}{(2\pi)^3} \int \delta[\eta(k)]\, d^3 k.$$  (3.16)

$k$ is the wave number vector which may be used for a classification of the one-electron states in the conduction band. For free electrons, it is

$$\eta(k) = \frac{k^2}{2m} - \mu.$$  (3.17)

Using this expression we may derive Eq. (3.15) from Eq. (3.16).

In order to prove our assertions we first regard electrons enclosed in a finite box of volume $\Omega$ with periodic boundary conditions. The one-electron wave functions may be classified by the wave number, or momentum, vector, $k$. But the periodic boundary conditions admit only discrete values of $k$. They form a lattice in $k$ space with a unit cell of volume $(2\pi)^3/\Omega$. The number of one-electron states in a thin layer of thickness $dk$ about the Fermi sphere of a radius equal to the Fermi momentum,

$$k_F = mv,$$  (3.18)

may be expressed in two ways,

$$\mathcal{N} \, d\eta = \frac{\Omega}{(2\pi)^3} \, 4\pi k_F^2 \, dk, \tag{3.19}$$

where use has been made of the verbal definition of $\mathcal{N}$. With

$$d\eta = v \, dk, \tag{3.20}$$

which follows from Eqs. (3.17) and (3.18), we obtain

$$\frac{\mathcal{N}}{\Omega} = \frac{m^2 v}{2\pi^2}. \tag{3.21}$$

If

$$|w_k(r')|^2 = \frac{1}{\Omega} = \overline{|w_k(r')|^2} \tag{3.22}$$

is remembered, Eq. (3.14) immediate leads to Eq. (3.15).

In the presence of a periodic potential, electron bands turn up. But the wave number vector may still be used to classify the one-electron states in a given band, in particular in the conduction band. Similar considerations as made above lead to

$$\mathcal{N} = \frac{\Omega}{(2\pi)^3} \int \delta[\eta(k)] \, d^3k. \tag{3.23}$$

We shall see in Section 8 that in this case $K_\omega(r, r')$ has to be replaced by another quantity, $\mathcal{K}_\omega(r, r')$, which is obtained from $K_\omega(r, r')$ by performing averages with respect to both position variables over a unit cell of the lattice. Therefore, this average has to be performed also on the righthand side of Eq. (3.14). Though Eq. (3.22) is in general not correct for one-electron wave functions in a periodic lattice, it does hold if an average is performed over a lattice cell. In this way, Eq. (3.16) is obtained.

If impurity atoms are introduced into the considerations and modeled as scattering potentials, the one-electron wave functions become much more complicated. Eqs. (2.21), (3.1) (3.2), and (3.6) are, however, still true since they do not depend upon the detailed structure of the one-electron Hamiltonian. They remain true after an averaging procedure has been performed with respect to the probability distribution of the positions of the impurity atoms. If the concentration of impurities is sufficiently low, both the impurity averaged number of states per energy interval near the Fermi energy and the impurity and position averaged $\overline{|w_m(r')|^2}$ are expected not to change much so that Eqs. (3.13), (3.15), and (3.16) still hold.

Some care is recommended regarding the definition of the density of states. In the theory of superconductivity it is customary to count only the orbital states, i.e. the number of states with one sign of the spin component. We followed this convention in the discussion presented above. In the theory of the electronic specific heat, however, the total number of states (with both signs of the spin component) is usually counted. If the coefficient, $\gamma$, of the electronic specific heat per unit volume is expressed by the above used definition of $N$, it is

$$\gamma = \frac{2\pi^2}{3} k_B^2 N , \tag{3.24}$$

where the Boltzmann constant, $k_B$, has explicitly been exhibited. There appears a renormalization factor on the righthand side of Eq. (3.24) for strong electron-phonon interaction which in the present discussion shall, however, be put equal to one.

The density of states is position independent in a homogeneous sample. Therefore, no position variable has been written down in Eqs. (3.15) and (3.16). The density of states does, however, become position dependent if the sample consists of several metals in electric contact. If the Fermi energy in two such metals is different, an electric field will be built up in a thin layer about the common interface so that the difference of the electric potential in the two metals compensates the difference in Fermi energy. The chemical potential becomes position independent in this way as is required in thermodynamic equilibrium. Apart from the immediate vicinity of the interface (distance of the order of the Fermi wavelength or less), it will be safe to replace $N(r)$ by its value in the respective homogeneous material.

### c) Remark on Strong-Coupling Superconductors

Only weak-coupling superconductors shall be treated in this review. It is, however, possible to extend the formulation to strong-coupling superconductors in which the frequency dependence of the phonon-mediated electron-electron interaction has to be taken into account in a more realistic way [26]. The pair potential becomes an $\omega$ dependent quantity, $\Delta_\omega(r)$. Eq. (2.15) is replaced by

$$\Delta_\omega(r) = - T \sum_{\omega'} \lambda(\omega, \omega') \int G_{\omega'}^{(0)}(r, r') \Delta_{\omega'}(r') G_{\omega'}^{(0)*}(r', r) \, d^3 r' . \tag{3.25}$$

The real and symmetric quantity $\lambda(\omega, \omega')$ contains the phonon-mediated electron-electron interaction and the shielded Coulomb repulsion. Eq. (1.23) for the Green's function of normal electrons is replaced by

$$(i\omega Z_n(\omega) + \mu) G_\omega^{(0)}(r, r') - \int d^3 r'' \langle r|h| r'' \rangle G_\omega^{(0)}(r'', r') = \delta(r - r') , \tag{3.26}$$

which follows from Eq. (1.23) by the simple replacement

$$\omega \to \omega Z_n(\omega) . \tag{3.27}$$

Here, the real quantity $Z_n(\omega)$ is defined by

$$\omega(1 - Z_n(\omega)) = \pi N T \sum_{\omega'} \lambda(\omega, \omega') \, \mathrm{sgn}\,\omega' , \tag{3.28}$$

where $N$ is the density of states without phonon enhancement. If the replacement (3.27) is made, the analysis given in this review [Eq. (5.8), e.g.] is easily extended to the strong-coupling case.

## 4. Calculation of the Kernel from Classical Mechanics

### a) Quantum-Mechanical and Classical Correlation Function

The Green's function $G_\omega^{(0)}(r, r')$ may in a certain sense be regarded as a quantum mechanical probability amplitude for the propagation of an electron from the position $r'$ to the position $r$. In the same spirit, the kernel $K_\omega(r, r')$, which according to Eq. (2.16) equals a product of two Green's functions, may be interpreted as a probability amplitude for the propagation of an electron pair. If the one-electron Hamiltonian is invariant under time reversal, Eq. (3.6) holds. The probability amplitude for the propagation of an electron pair becomes equal to the probability itself, of the propagation of an electron. It will turn out in the following that this probability may even be calculated on the basis of classical mechanics instead of quantum mechanics.

The detailed formulation may be based upon Eq. (3.11), which is valid provided that $\langle r|h|r'\rangle$ is invariant under time reversal [Eq. (3.3)]. Eq. (3.11) may evidently be written in the form

$$K_\omega(r, r') = i \int_0^\infty dt \sum_n \frac{\exp[-(i\eta_n + |\omega|)t]}{\eta_n + i|\omega|} \mathscr{F}_n(r, r'; t) \tag{4.1}$$

with $\mathscr{F}_n(r, r'; t)$ given by

$$\mathscr{F}_n(r, r'; t) := \sum_m w_n^*(r) w_m(r) w_m^*(r') w_n(r') \exp[i(\eta_n - \eta_m)t]. \tag{4.2}$$

Introducing time dependent position operators, $r(t)$ and $r(0)$, we see that Eq. (4.2) may also be written as

$$\mathscr{F}_n(r, r'; t) = (w_n, \delta(r(t) - r) \delta(r(0) - r') w_n). \tag{4.3}$$

The righthand side may be interpreted as a time dependent quantum-mechanical correlation function [1].

It is suggestive to try an evaluation of the sum on the righthand side of Eq. (4.1) for not too small samples in the same way as this has been done in Eq. (3.12). If $\mathscr{F}_n(r, r'; t)$ or, rather, $\overline{\mathscr{F}_n(r, r'; t)}$ [averaged over all states with $\eta \approx \eta_n$; cf. Eq. (3.14)] were a slowly varying function of the variable $\eta$, the result would be [1]

$$K_\omega(r, r') = 2\pi \int_0^\infty dt \exp[-2|\omega|t] \mathscr{F}(r, r'; t) \tag{4.4}$$

with

$$\mathscr{F}(r, r'; t) := \mathscr{N} \overline{\mathscr{F}_n(r, r'; t)}. \tag{4.5}$$

Here, the average has to be performed on the Fermi surface ($\eta = 0$). Eq. (4.4) states that the kernel, $K_\omega(r, r')$, is the Laplace transform of the correlation function (4.5).

The assumption leading to Eq. (4.4) is, however, not correct. The analysis of a simple example [2] shows that the righthand side of Eq. (4.5) is certainly not a slowly varying function of $\eta$. A reasonable interpretation can be given to Eq. (4.4) only if it is assumed that $\mathscr{F}(r, r'; t)$ is not equal to the righthand side of Eq. (4.5) but rather to its slowly varying part. It might now be conjectured that this slowly varying part is given by its classical analogue which is defined as the average on the properly normalized micro-canocical ensemble,

$$\mathscr{F}(r, r'; t) := \frac{1}{(2\pi)^3} \int \delta[\eta(s', k')] \, \delta[R(s', k'; t) - r] \, \delta(s' - r') \, d^3 s' \, d^3 k'. \tag{4.6}$$

The symbols $s'$ and $k'$ are position and canonical momentum variables, respectively. $\eta$, the difference between energy and chemical potential, is regarded as a function of both position and momentum in order to make the formulation sufficiently general. $R(s', k'; t)$ denotes the position, at time $t$, of an electron which has started, at $t = 0$, from the position $s'$ with momentum $k'$. From now on, the symbol $\mathscr{F}(r, r'; t)$ shall only be used for the classical correlation function defined by Eq. (4.6). $\mathscr{F}(r, r'; t)$ is different from zero, and actually infinite, whenever the micro-canonical ensemble contains an orbit on which an electron may pass through the position $r'$ at time zero and through $r$ at time $t$.

We check the classical formulation on the example of free electrons in an infinite volume. In the absence of a magnetic field (invariance under time reversal!) Eq. (2.25) reduces to

$$\left(i\omega + \mu + \frac{1}{2m} \frac{\partial^2}{\partial r^2}\right) G_\omega^{(0)}(r, r') = \delta(r - r'). \tag{4.7}$$

The unique solution which is bounded, and actually vanishes, at infinity is given by

$$G_\omega^{(0)}(r, r') = -\frac{m}{2\pi|r - r'|} \exp[i\sqrt{2m(\mu + i\omega)} \, |r - r'|], \tag{4.8}$$

where the square root with positive imaginary part has to be selected. For

$$|\omega| \ll \mu \tag{4.9}$$

this equation reduces to

$$G_\omega^{(0)}(r, r') = -\frac{m}{2\pi|r - r'|} \exp\left[\left(i m v \operatorname{sgn} \omega - \frac{|\omega|}{v}\right)|r - r'|\right], \tag{4.10}$$

where $v$ is the Fermi velocity. By virtue of Eq. (2.16), $K_\omega(r, r')$ is then given by

$$K_\omega(r, r') = \left(\frac{m}{2\pi|r-r'|}\right)^2 \exp\left[-\frac{2|\omega| |r-r'|}{v}\right].\qquad(4.11)$$

No oscillatory part occurs on the righthand side.

In the corresponding classical problem it is

$$\eta(s', k') = \eta(k') = \frac{k'^2}{2m} - \mu\qquad(4.12)$$

and

$$R(s', k'; t) = s' + k'\frac{t}{m}.\qquad(4.13)$$

If now the delta function, $\hat{\delta}(a, b)$, of the directions of the vectors $a$ and $b$ is defined by

$$\hat{\delta}(a, b) = \hat{\delta}(b, a) \neq 0 \quad \text{only if } a \text{ and } b \text{ parallel,}$$
$$\oint \hat{\delta}(a, b) \, d\Omega_a = 1,\qquad(4.14)$$

(with integration over all directions of $a$) we may write

$$\delta[R(s', k'; t) - r] = \frac{1}{|r-s'|^2} \delta\left(|r-s'| - \frac{|k'|}{m}t\right)\hat{\delta}(r-s', k')\quad(4.15)$$

for $t > 0$. The integration with respect to $s'$ may, as always, immediately be performed in Eq. (4.6). Also carrying out the $k'$ integration with respect to direction and modulus, we finally arrive at

$$\mathscr{F}(r, r'; t) = \frac{m^2 v}{(2\pi)^3} \frac{\delta(|r-r'|-vt)}{|r-r'|^2},\qquad(4.16)$$

where use has been made of Eq. (4.14) and of

$$\int_0^\infty \delta[\eta(k')] k'^2 \, dk' = m^2 v.\qquad(4.17)$$

Inserting Eq. (4.16) into Eq. (4.4), we again obtain Eq. (4.11). The classical formulation is, therefore, confirmed in the case of free electrons under the important additional assumption (4.9).

The physical ideas that led us to Eqs. (4.4) and (4.6) will be further developed below in this section and in Section 6. We shall finally arrive at a formulation which essentially consists of a Boltzmann equation plus boundary conditions. The Boltzmann equation will be justified on a quantum-mechanical basis in Sections 7 and 8 so that no conjectures are required for this part of the formulation. Only the boundary conditions will still have to be based upon physical intuition.

## b) General Properties of the Kernel Confirmed

We want to check whether the general properties of the kernel, $K_\omega(r, r')$, as summarized and proved on a quantum-mechanical basis in Section 3, also hold in the classical formulation as defined by Eqs. (4.4) and (4.6).

Eq. (3.1) follows immediately since the righthand side of Eq. (4.4) contains only the modulus of $\omega$. Further, $K_\omega(r, r')$ is certainly non-negative [Eq. (3.6)] since $\mathscr{F}(r, r'; t)$ is so and since the righthand side of Eq. (4.4) is an integral with a non-negative kernel. The symmetry in the arguments,

$$K_\omega(r, r') = K_\omega(r', r) \tag{4.18}$$

[Eq. (3.6)], follows since, in a system invariant under time reversal, all electron orbits may be reversed in time.

We give a complete formal proof of the last statement. If it is

$$R(s', k'; t) = s \tag{4.19}$$

and if this electron arrives in the position $s$ with momentum $k$ we may also write

$$R(s, k; -t) = s'. \tag{4.20}$$

The conservation of energy may be stated in the form

$$\eta(s', k') = \eta(s, k). \tag{4.21}$$

Finally it is

$$d^3 s' \, d^3 k' = d^3 s \, d^3 k \tag{4.22}$$

according to Liouville's theorem. Inserting all this into Eq. (4.6) we obtain

$$\mathscr{F}(r, r'; t) = \frac{1}{(2\pi)^3} \int \delta[\eta(s, k)] \, \delta(s - r) \, \delta[R(s, k; -t) - r'] \, d^3 s \, d^3 k, \tag{4.23}$$

where so far no use has been made of the invariance under time reversal. It follows, however, from this invariance that it is

$$\eta(s, k) = \eta(s, -k), \tag{4.24}$$

and that all orbits may be reversed,

$$R(s, k; -t) = R(s, -k; t). \tag{4.25}$$

Actually, Eq. (4.25) follows from Eq. (4.24) with the help of Hamilton's equations of motion. Eq. (4.23) may now be written in the form

$$\mathscr{F}(r, r'; t) = \frac{1}{(2\pi)^3} \int \delta[\eta(s, -k)] \, \delta(s - r) \, \delta[R(s, -k; t) - r'] \, d^3 s \, d^3 k. \tag{4.26}$$

Comparison with Eq. (4.6) leads to

$$\mathscr{F}(r', r; t) = \mathscr{F}(r, r'; t) \tag{4.27}$$

from which relation Eq. (4.18) is immediately obtained.

The sum rule (3.13) is also easily derived. First it follows from Eq. (4.6) that it is

$$\int \mathscr{F}(r, r'; t)\, d^3 r = N(r'),\qquad (4.28)$$

where the righthand side does not depend upon time. Use has been made of the general definition (3.16) of the density of states. Application of Eq. (4.4) leads to Eq. (3.13).

### c) Distribution Function in Phase Space

Eq. (4.6) or, rather, Eq. (4.23) may be brought into a more convenient form by carrying out the space integration in the latter. The result is

$$\mathscr{F}(r, r'; t) = \int f(r, k; r'; t)\, d^3 k \qquad (4.29)$$

with

$$f(r, k; r'; t) = \frac{1}{(2\pi)^3}\, \delta[\eta(r, k)]\, \delta[R(r, k; -t) - r'].\qquad (4.30)$$

Defining a function, $g_\omega(r, k; r')$, by

$$g_\omega(r, k; r')\, \delta[\eta(r, k)] := 2\pi \int_0^\infty \exp[-2|\omega|\, t]\, f(r, k; r'; t)\, dt,\quad (4.31)$$

we finally obtain [2]

$$K_\omega(r, r') = \int g_\omega(r, k; r')\, \delta[\eta(r, k)]\, d^3 k \qquad (4.32)$$

from Eqs. (4.4), (4.29), and (4.31). Eq. (4.32) plays a central role in the work to be reported on in this review article.

We claim that $f(r, k; r'; t)$ as defined by Eq. (4.30) is, with respect to the variables $r$ and $k$, a distribution function of classical electrons in phase space. Because of Liouville's theorem it suffices to show that $f(r(t), k(t); r'; t)$ is time independent, if $r(t)$ and $k(t)$ are time dependent position and momentum vectors of a classical electron moving with Fermi energy ($\eta = 0$). This assertion follows immediately from Eq. (4.30) since both $\eta(r(t), k(t))$ and $R(r(t), k(t); -t)$ (the starting point) are constant along each orbit. For $t = 0$, Eq. (4.30) reduces to

$$f(r, k; r'; 0) = \frac{1}{(2\pi)^3}\, \delta[\eta(r, k)]\, \delta(r - r') \qquad (4.33)$$

so that $f(r, k; r'; t)$ is a distribution function of classical particles at time $t$ which, at $t = 0$, started from the position $r'$ with Fermi energy and isotropic momentum distribution*.

---

* This formulation is literally correct for spherical Fermi surfaces only. In general, the momentum distribution at the start is that of the micro-canonical ensemble.

The distribution function $g_\omega(r, k; r')$ is defined for $k$ on the Fermi surface only; cf. Eq. (4.31). It may be regarded as the (with respect to time) Laplace transformed distribution function of electrons having started from $r'$ with Fermi energy and isotropic momentum distribution. According to Eq. (4.32), $K_\omega(r, r')$ may then, with respect to the variable $r$, be interpreted as the corresponding particle density in real space. The linearized gap equation (2.15) states that the classical electrons moving from $r'$ to $r$ in a way transport information as to the value of the pair potential at the point $r'$ to the point $r$.

The explicit calculation of the distribution function may easily be done for free electrons. It seems convenient to replace the momentum, $k$, by the velocity, $v$, according to

$$k = mv. \tag{4.34}$$

Eq. (4.30), with

$$R(r, k; -t) = r - vt, \tag{4.35}$$

leads to

$$f(r, k; r'; t) = \frac{1}{(2\pi)^3 |r - r'|^2} \delta[\eta(r, k)] \hat{\delta}(r - r', v) \delta(|r - r'| - vt), \tag{4.36}$$

where use has been made of Eq. (4.15). With the help of

$$\delta(|r - r'| - vt) = \frac{1}{v} \delta\left(t - \frac{|r - r'|}{v}\right) \tag{4.37}$$

and of Eq. (4.31), $g_\omega(r, k; r')$ is calculated as

$$g_\omega(r, k; r') = \frac{1}{(2\pi|r - r'|)^2 v} \exp\left[-\frac{2|\omega| \, |r - r'|}{v}\right] \hat{\delta}(r - r', v). \tag{4.38}$$

Eq. (4.32) reads

$$K_\omega(r, r') = m^2 v \oint g_\omega(r, k; r') \, d\Omega \tag{4.39}$$

for free electrons. $v$ is again the Fermi velocity. The integration is done with respect to all directions of the vector $k$ or the vector $v$. Inserting Eq. (4.38) and making use of Eq. (4.14) we are led to Eq. (4.11) which we already know to be the result of a quantum-mechanical calculation provided that Eq. (4.9) holds.

## d) Electrons in a Crystal

The conduction electrons in a metal are not free. The appearence of energy bands is a consequence of quantum mechanics. It is, nevertheless, possible to formulate the motion of the conduction electrons in the language of classical mechanics. We shall first give a survey of this classical formulation and afterwards apply it to a calculation of the kernel, $K_\omega(r, r')$.

Let the electronic energy (minus Fermi energy) in the conduction band [i.e., the eigenvalue of the Schrödinger equation (3.7)] be given as a function of the wave number, $k$, $\eta = \eta(k)$. It may then be shown that the electronic velocity is

$$v(k) := \frac{\partial \eta}{\partial k} . \tag{4.40}$$

This equation may also be interpreted as a canonical equation of motion with Hamiltonian $\eta(k)$ so that the classical concept of a distribution function in phase space is still applicable. The second derivatives of $\eta$ with respect to the Cartesian components of the momentum vector define a symmetric tensor,

$$\left(\frac{1}{m}\right)_{ij} := \frac{\partial^2 \eta}{\partial k_i \partial k_j} . \tag{4.41}$$

The inverse tensor, $m_{ij}$, is called the mass tensor.

For the special case of spherical energy surfaces,

$$\eta(k) = f\left(\frac{k^2}{2}\right), \tag{4.42}$$

velocity and momentum are parallel to each other,

$$v = f' k . \tag{4.43}$$

Here, $f'$ is the derivative of $f$ [Eq. (4.42)] with respect to its argument. An effective mass, $m^*$, is introduced for spherical energy surfaces by

$$k = : m^* v ; \tag{4.44}$$

cf. Eq. (4.34). The effective mass will in general depend upon the energy but it will later be needed at the Fermi energy ($\eta = 0$) only. Comparison of Eqs. (4.44) and (4.43) shows that it is

$$m^* = \frac{1}{f'} . \tag{4.45}$$

We summarize a few relations which will be needed later on,

$$\left(\frac{1}{m}\right)_{ij} = f'\delta_{ij} + f''k_i k_j , \tag{4.46}$$

$$m_{ij} = m^*\delta_{ij} - \frac{m^{*2} f''}{1 + k^2 m^* f''} k_i k_j , \tag{4.47}$$

$$\det(m) = \frac{m^{*3}}{1 + k^2 m^* f''} . \tag{4.48}$$

Eq. (4.46) follows immediately from Eqs. (4.41) and (4.42). Eq. (4.47) is then obtained by calculating the inverse tensor. $\det(m)$ finally denotes the determinant of the mass tensor.

Returning now to the general case we easily see that it is

$$d^3k = \frac{1}{v(k)}\,dS\,d\eta\,. \tag{4.49}$$

Here, $dS$ is the surface element on a surface of constant energy and $v(k)$ is the modulus of the velocity vector introduced by Eq. (4.40). Making use of Eq. (4.49) we obtain

$$K_\omega(r, r') = \oint g_\omega(r, k; r')\,\frac{dS}{v(k)} \tag{4.50}$$

from Eq. (4.32), and

$$N = \frac{1}{(2\pi)^3}\oint \frac{dS}{v(k)} \tag{4.51}$$

from Eq. (3.16). The integrations are extended over the Fermi surface with surface element $dS$ and velocity $v(k)$. Eq. (4.38) is obviously replaced by

$$g_\omega(r, k; r') = \frac{1}{(2\pi|r - r'|)^2\,v(k)}\exp\left[-\frac{2|\omega|\,|r - r'|}{v(k)}\right]\hat{\delta}(r - r', v(k))\,. \tag{4.52}$$

In the special case of spherical energy surfaces it is

$$dS = k^2 d\Omega\,. \tag{4.53}$$

It then follows from Eqs. (4.51) and (4.44) that it is

$$N = \frac{m^{*2}v}{2\pi^2}\,. \tag{4.54}$$

This equation closely resembles Eq. (3.15). Similarly, Eq. (4.50) may be written as

$$K_\omega(r, r') = m^{*2}v\oint g_\omega(r, k; r')\,d\Omega\,, \tag{4.55}$$

which is analogous to Eq. (4.39). Inserting Eq. (4.52) we finally obtain a result which is identical with Eq. (4.11) apart from a replacement of the electronic mass, $m$, by the effective mass, $m^*$. From now on we shall use the simpler phrase "spherical Fermi surface" instead of "spherical energy surfaces in the neighbourhood of the Fermi surface".

The case of an arbitrary Fermi surface is more intricate though Eq. (4.52) is, of course, still valid. A calculation presented below in small

3*

print leads to

$$K_\omega(r, r') = \frac{1}{(2\pi)^2} \left| \frac{\det(m(k))}{\sum\limits_{ij} m_{ij}(k)(x_i - x_i')(x_j - x_j')} \right| \exp\left[ -\frac{2|\omega|\,|r - r'|}{v(k)} \right]. \quad (4.56)$$

Both the mass tensor and the velocity vector depend upon the momentum, $k$, which has to be chosen in such a way that $v(k)$ is parallel to $r - r'$. If this condition is satisfied for several values of $k$ on the Fermi surface, the righthand side of Eq. (4.56) has to be replaced by a sum of several such contributions. The treatment is not quite unproblematic since vectors $k$ in an infinitesimal neighbourhood of each other may lead to parallel vectors $v(k)$. In such a case, the denominator in Eq. (4.56) vanishes. We shall, however, not study this complication. In the special case of a spherical Fermi surface, application of Eqs. (4.47) and (4.48) again leads from Eq. (4.56) to Eq. (4.11) with $m$ replaced by $m^*$. The quantum mechanical proof of Eq. (4.56) is postponed to Section 8.

For the derivation of Eq. (4.56) from Eqs. (4.50) and (4.52) we first assume a special orientation of the Cartesian coordinate system so that $r - r'$ is parallel to the $z$ direction. If $k$ is varied on the Fermi surface in the neighbourhood of the the point leading to a $v(k)$ parallel to $r - r'$ we obtain

$$\begin{pmatrix} dv_x \\ dv_y \end{pmatrix} \begin{pmatrix} \left(\dfrac{1}{m}\right)_{xx} & \left(\dfrac{1}{m}\right)_{xy} \\ \left(\dfrac{1}{m}\right)_{yx} & \left(\dfrac{1}{m}\right)_{yy} \end{pmatrix} \begin{pmatrix} dk_x \\ dk_y \end{pmatrix} \quad (4.57)$$

from Eqs. (4.40) and (4.41). The terms proportional to $dk_z$ have not been written down since this infinitesimal quantity is small of the second order if the surface normal points into the $z$ direction. It follows from Eq. (4.57) that it is

$$\frac{v^2 d\Omega}{dS} = \left| \left(\frac{1}{m}\right)_{xx} \left(\frac{1}{m}\right)_{yy} - \left(\frac{1}{m}\right)_{xy} \left(\frac{1}{m}\right)_{yx} \right|, \quad (4.58)$$

where $d\Omega$ is the infinitesimal solid angle of the directions of $v$, and $dS$ has been explained in connection with Eq. (4.48). The righthand side may be expressed by the mass tensor itself,

$$\left(\frac{1}{m}\right)_{xx} \left(\frac{1}{m}\right)_{yy} - \left(\frac{1}{m}\right)_{xy} \left(\frac{1}{m}\right)_{yx} = \frac{m_{zz}}{\det(m)}. \quad (4.59)$$

If the orientation of the coordinate system is now chosen in an arbitrary way it follows from the last two equations that it is

$$\frac{v^2 d\Omega}{dS} = \left| \frac{\sum\limits_{ij} m_{ij}(x_i - x_i')(x_j - x_j')}{\det(m)\,|r - r'|^2} \right|. \quad (4.60)$$

Eq. (4.56) is an immediate consequence of this equation and of Eqs. (4.50) and (4.52).

## 5. Boltzmann Equation

### a) General Formulations

If a metal contains impurity atoms it would be an impossible and perhaps even meaningless task to calculate the classical orbits and the distribution functions, introduced in the previous section, for the actual distribution of impurity atoms. A manageable task of physical interest is to do calculations for a statistical distribution of impurity atoms. It is expected that in this case the distribution function $f(r, k; r'; t)$ will satisfy a Boltzmann equation, which we are now going to derive in a heuristic manner.

As pointed out in Section 4c, the distribution function $f(r, k; r'; t)$ does not depend upon time if $r$ and $k$ are replaced by time dependent position and momentum variables, $r(t)$ and $k(t)$, of a classical particle. Therefore, we may write

$$\frac{d}{dt} f(r(t), k(t); r'; t) = \left( \frac{\partial}{\partial t} + \dot{r} \cdot \frac{\partial}{\partial r} + \dot{k} \cdot \frac{\partial}{\partial k} \right) f(r, k; r'; t) = 0, \quad (5.1)$$

with

$$\dot{r} = \frac{\partial \eta(r, k)}{\partial k}, \qquad \dot{k} = -\frac{\partial \eta(r, k)}{\partial r}. \quad (5.2)$$

Eqs. (5.2) are Hamilton's canonical equations of motion. Eq. (5.1) is sometimes called Liouville's equation. The micro-canonical ensemble,

$$f(r, k) = \frac{1}{(2\pi)^3} \delta[\eta(r, k)], \quad (5.3)$$

is a stationary (i.e., time independent) solution to Eq. (5.1).

Now it is assumed that $\eta(r, k)$ does not contain the impurity atoms which shall be introduced afterwards in a statistical manner. By an elastic collision with an impurity atom the electron is deflected. Its momentum vector, $k$, is altered with the energy, $\eta(r, k)$, remaining constant. Therefore, the distribution function $f(r(t), k(t); r'; t)$ changes in time due to scattering processes so that the righthand side of Eq. (5.1) is no longer equal to zero. Eq. (5.1) has to be replaced by

$$\frac{d}{dt} f(r(t); k(t); r'; t) = -n v(k) \sigma(k) f(r, k; r'; t)$$
$$+ n \int P(k, k') \delta[\eta(k) - \eta(k')] f(r, k'; r'; t) d^3k', \quad (5.4)$$

where the first term on the righthand side represents electrons which had the momentum $k$ before the collision process and the second one those which obtain the momentum $k$ by the collision. In postulating Eq. (5.4) it has tacitely been assumed that the collision time and the range of the interaction are sufficiently short. It is definitely not required that the collision process itself can be described in terms of classical mechanics.

Eq. (5.4) shall be applied to the interior of a homogeneous metal. Therefore, the energy, $\eta$, does not depend upon the position variable, $r$. In Eq. (5.4), $v(k)$ is the modulus of the velocity vector, $v(k)$, defined by Eq. (4.40). $n$ is the concentration (number per volume) of impurity atoms. $\sigma(k)$ is called the scattering cross section. It is related to the scattering probability, $P(k', k)$, by

$$v(k)\,\sigma(k) = \int P(k', k)\,\delta\,[\eta(k) - \eta(k')]\,d^3 k'. \tag{5.5}$$

Because of their physical meaning, both $\sigma(k)$ and $P(k', k)$ are non-negative quantities,

$$P(k', k) \geqq 0, \qquad \sigma(k) \geqq 0. \tag{5.6}$$

The second relation obviously is a consequence of the first one.

Combining Eq. (5.4) with Eq. (5.1) we obtain the Boltzmann equation

$$\left(\frac{\partial}{\partial t} + v(k) \cdot \frac{\partial}{\partial r} + n v(k)\,\sigma(k)\right) f(r, k; r'; t)$$
$$- n \int P(k, k')\,\delta\,[\eta(k) - \eta(k')]\,f(r, k'; r'; t)\,d^3 k' = 0. \tag{5.7}$$

The force term has been dropped since it vanishes inside a homogeneous metal. For $\eta(k) = 0$ the last term in Eq. (5.7) may also be written as an integral over the Fermi surface; cf. Eqs. (4.50) and (4.51). The initial condition is still given by Eq. (4.33). If the Laplace transformed distribution function $g_\omega(r, k; r')$ is introduced with the help of Eq. (4.31), we arrive at a Laplace transformed Boltzmann equation,

$$\left(2|\omega| + v(k) \cdot \frac{\partial}{\partial r} + n v(k)\,\sigma(k)\right) g_\omega(r, k; r')$$
$$- n \int P(k, k')\,\delta\,[\eta(k')]\,g_\omega(r, k'; r')\,d^3 k' = \frac{1}{(2\pi)^2}\,\delta(r - r'), \tag{5.8}$$

where the inhomogeneity on the righthand side follows from the initial condition. The Boltzmann equation (5.8) plays a central role in the method of the correlation function. It shall be proved on a microscopic quantum-mechanical basis in Sections 7 (free electrons) and 8 (electrons in a crystal). Eq. (5.8) has to be supplemented by boundary conditions on surfaces of the sample and on interfaces between different metals. These boundary conditions shall be formulated in Section 6.

For free electrons or electrons with spherical Fermi surface, momentum vector, $k$, and velocity vector, $v$, are parallel to each other. The momentum variable in the distribution functions may be replaced by the velocity which, perhaps, appeals more to physical intuition. The scattering probability, $P(k, k')$, may be expressed by the non-negative differential scattering cross section,

$$\frac{d\sigma(v, v')}{d\Omega} := m^{*2} P(k, k'),\qquad (5.9)$$

for a scattering process leading from velocity $v'$ to velocity $v$. The modulus of both velocity vectors equals the Fermi velocity, $v$. The symbol $m^*$ denotes the effective mass of the electrons on the Fermi sphere as introduced by Eq. (4.44). It has to be replaced by the electronic mass, $m$, for free electrons. Eq. (5.5) may be written as

$$\sigma(v) = \oint \frac{d\sigma(v', v)}{d\Omega} d\Omega' \qquad (5.10)$$

with integration over all directions of the vector $v'$. This equation is the usual relation between the scattering cross section and the differential scattering cross section. With the help of Eq. (5.9), Eq. (5.8) may now be written

$$\left(2|\omega| + v \cdot \frac{\partial}{\partial r} + nv\sigma(v)\right) g_\omega(r, v; r')$$

$$- nv\oint \frac{d\sigma(v, v')}{d\Omega} g_\omega(r, v'; r') d\Omega' = \frac{1}{(2\pi)^2} \delta(r - r'). \qquad (5.11)$$

The integration over the Fermi sphere has been replaced by an integration over all directions of $v'$.

For free electrons, the differential scattering cross section usually depends only upon the angle, $\theta$, between the vectors $v$ and $v'$,

$$\frac{d\sigma(v, v')}{d\Omega} = \frac{d\sigma(\theta)}{d\Omega}. \qquad (5.12)$$

This assumption will in the following mostly be made also for electrons with spherical Fermi surface. Under the assumption (5.12), the scattering cross section,

$$\sigma = \oint \frac{d\sigma(\theta)}{d\Omega} d\Omega, \qquad (5.13)$$

does not depend upon the direction of the velocity vector. We write down the Boltzmann equation under these special assumptions [2],

$$\left(2|\omega| + v \cdot \frac{\partial}{\partial r} + nv\sigma\right) g_\omega(r, v; r')$$

$$- nv\oint \frac{d\sigma(\theta)}{d\Omega} g_\omega(r, v'; r') d\Omega' = \frac{1}{(2\pi)^2} \delta(r - r'). \qquad (5.14)$$

An assumption basic for the present discussion was that of invariance under time reversal. The Boltzmann equation derived so far is in particular not valid in the presence of a magnetic field. It would certainly not suffice to start from the Hamiltonian formulation of an electron moving in a magnetic field and to include a force term, corresponding to the Lorentz force, in the Boltzmann equation since the relation between $K_\omega(r, r')$ and classical particles breaks down in this case.

It is, however, possible to conjecture [7] the correct generalization of the Boltzmann equation to the case of an external magnetic field. The behaviour of $K_\omega(r, r')$ under gauge transformations, Eq. (1.27), has been stated in Eq. (3.2), which was obtained from the quantum-mechanical definition of $K_\omega(r, r')$. It seems a nearlying assumption that the analogous relation holds for the distribution function,

$$g_\omega(r, k; r') \to g'_\omega(r, k; r') = g_\omega(r, k; r') \exp[2ie(U(r') - U(r))],$$
$$g_\omega(r, v; r') \to g'_\omega(r, v; r') = g_\omega(r, v; r') \exp[2ie(U(r') - U(r))].$$

$$(5.15)$$

This gauge transformation is compatible with the Boltzmann equation if the gradient, $\partial/\partial r$, is replaced by an expression known from quantum mechanics of a charged particle,

$$\frac{\partial}{\partial r} \to \tilde{\partial} = \frac{\partial}{\partial r} + 2ieA(r),$$

$$(5.16)$$

where $-e$ is the charge of an electron and $A(r)$ is a vector potential of the external magnetic field [Eq. (1.24)]. Remarkably enough, twice the electronic charge occurs in Eq. (5.16). The physical reason is that the carriers of charge in superconductors are electron pairs.

The magnetic field acting on the electrons is identical with the external field since the electric current in the superconductor vanishes at a transition of the second kind [$A(r) \to 0$]. It will be a result of Sections 7 and 8 that no Lorentz force term has to be introduced into the Boltzmann equation. The curvature of the orbits due to the magnetic field has definitely to be neglected.

After the replacement (5.16) has been made, the Boltzmann equation no longer admits real solutions. The function $g_\omega(r, k; r')$ can no longer be interpreted as a distribution function of particles. This was to be expected because of the breakdown of time reversal invariance. In order to still have a somewhat suggestive terminology, we shall speak of a distribution function, not of particles, but of carriers of information. The gap equation (2.15) may now be interpreted as stating that these carriers transport information as to the value of the pair potential from the point $r'$ to the point $r$. Even the orbits of such carriers of information remain a meaningful concept as will turn out in Section 12.

## b) Restrictions on Scattering Probability and Cross Sections. General Properties of the Kernel Confirmed

There are restrictions on the scattering probability, $P(k, k')$, the differential scattering cross section, $d\sigma(v, v')/d\Omega$, and the scattering cross section, $\sigma(k)$ or $\sigma(v)$, which follow from general principles. Similar restrictions will occur in the next section in connection with the formulation of boundary conditions. The general properties of the kernel, $K_\omega(r, r')$, as discussed in Section 3 will be shown to be consequences [2] of the restrictions just mentioned, provided that the kernel is not calculated on a quantum-mechanical basis but with the help of the method of the correlation function.

It shall be shown in Section 11a that Boltzmann equation plus boundary conditions determine the distribution function uniquely. Since the inhomogeneity on the righthand sides of Eqs. (5.8), (5.11), and (5.14) is real and since the coefficients on the lefthand sides are also real in the absence of a magnetic field $[A(r) \equiv 0]$, it follows from the uniqueness theorem just mentioned that also the distribution function and the kernel are real quantities in this case. In this way, part of the statement (3.6) is confirmed.

Also the non-negativeness of distribution function and kernel in the absence of a magnetic field can be proved. It is a consequence of the non-negativeness of the scattering probability (or differential scattering cross section) and the scattering cross section, Eq. (5.6).

We sketch a proof of the last statement by indicating that the time dependent distribution function, $f(r, k; r'; t)$ is never negative. It certainly is non-negative for $t = 0$ because of Eq. (4.33). Then we show that $f(r(t), k; r'; t)$ cannot become negative for $t > 0$. Here, $r(t)$ denotes the straight classical orbit of momentum $k$ without impurity scattering. $f(r(t), k; r'; t)$ would remain non-negative if the second term on the righthand side of Eq. (5.4) were discarded. But this second term always leads to an increase of $f(r(t), k; r'; t)$ because of the second Eq. (5.6). This discussion is correct only for orbits not hitting surfaces or interfaces. Otherwise the boundary conditions would have to be taken into account.

The classical motion is particle conserving. This has in particular been expressed by Eqs. (5.5), (5.10), and (5.13). We want to show that the sum rule (3.13) is a consequence of this conservation of particles. Integrating Eq. (5.7) over the whole $k$ space and making use of Eq. (5.5) we obtain

$$\int d^3k \left( \frac{\partial}{\partial t} + v(k) \cdot \frac{\partial}{\partial r} \right) f(r, k; r'; t) = 0. \tag{5.17}$$

Integrating this equation with respect to $r$ over the whole sample we derive the result that the total number of particles,

$$\mathcal{N}(t) := \int f(r, k; r'; t) d^3r \, d^3k, \tag{5.18}$$

is constant in time and equal to its initial value,

$$\mathcal{N}(t) = \mathcal{N}(0) = N(r'),$$ (5.19)

[cf. Eqs. (4.33) and (3.16)] provided that a certain surface and interface integral vanishes (in more physical terms: provided that no particles are created or destroyed on surfaces and interfaces). It then follows from Eqs. (4.4), (4.29), (5.18), and (5.19) that it is

$$\int K_\omega(r, r') \, d^3 r = 2\pi \int_0^\infty dt \exp(-2|\omega| t) N(r') = \frac{\pi N(r')}{|\omega|}.$$ (5.20)

This is the sum rule (3.13) to be proved.

The vanishing surface and interface integral mentioned above may be written in the form

$$\sum_j \int d^3 k \oint_{B_j} d^2 r \cdot v_j(k) f_j(r, k; r'; t) = 0.$$ (5.21)

The letter $j$ labels the various metals in electric contact which constitute the sample. $f_j(r, k; r'; t)$ is the distribution function and $v_j(k)$ the momentum dependent velocity vector in metal $j$. The $r$ integration has to be extended over the boundary, $B_j$, of each separate metal. The vectorial surface element, $d^2 r$, is parallel to the boundary normal and points towards the exterior of the particular metal. There is only one contribution to a point on the external surface but there are two contributions (one from each of the adjoining metals) to a point on an interface. It will turn out in Section 6 that the lefthand side of Eq. (5.21) vanishes even without the $r$ integration being performed. We have

$$\int d^3 k \, n \cdot v_j(k) f_j(r, k; v'; t) = 0$$ (5.22)

in an arbitray point, $r$, of an external surface. $n$ is the normal unit vector pointing out of the metal. The sum of two such terms with opposite normal vectors vanishes on an interface. The physical meaning of these conditions is that the normal component of the particle flux vanishes on external surfaces and is continuous on interfaces.

Introducing the distribution function $g_{j\omega}(r, k; r')$ with the help of Eq. (4.31), we obtain

$$\sum_j \int d^3 k \, \delta[\eta_j(k)] \oint_{B_j} d^2 r \cdot v_j(k) g_{j\omega}(r, k; r') = 0$$ (5.23)

from Eq. (5.21). $\eta_j(k)$ is the momentum dependent energy in metal $j$. For spherical Fermi surfaces we finally arrive at

$$\sum_j N_j \oint d\Omega \oint_{B_j} d^2 r \cdot v_j \, g_\omega(r, v_j; r') = 0$$ (5.24)

where $N_j$ is the density of states [Eq. (3.16)] in metal $j$. The label $j$ on the distribution function is now attached to the velocity vector. The density of states may, according to Eq. (4.54), be expressed in terms of the effective mass, $m_j^*$, of the conduction electrons and the Fermi velocity, $v_j$.

The postulate of time reversal invariance played a decisive role for the possibility of introducing a classical particle picture. Time reversal invariance of the scattering process means that it is

$$P(k, k') = P(-k', -k), \quad \sigma(k) = \sigma(-k).$$ (5.25)

If, for spherical Fermi surfaces, the differential scattering cross section is introduced with the help of Eq. (5.9), the Eqs. (5.25) may be written as

$$\frac{d\sigma(v, v')}{d\Omega} = \frac{d\sigma(-v', -v)}{d\Omega}; \quad \sigma(v) = \sigma(-v). \tag{5.26}$$

These relations are automatically fulfilled if the differential scattering cross sections depends only upon the angle, $\theta$, between the vectors $v$ and $v'$ [Eq. (5.12)]. Now we do not postulate that the magnetic field vanishes and want to show that the kernel, $K_\omega(r, r')$, is Hermitian as a consequence of time reversal invariance. This Hermiticity [Eq. (2.21)] is an almost trivial statement on the quantum-mechanical level.

We start from Eq. (5.8) with the replacement (5.16) made, or from

$$(2|\omega| + v(k) \cdot \tilde{\partial} + n v(k) \sigma(k)) g_\omega(r, k; r')$$
$$- n \int P(k, k') \delta[\eta(k')] g_\omega(r, k'; r') d^3 k' = \frac{1}{(2\pi)^2} \delta(r - r'). \tag{5.27}$$

This equation is now multiplied by $\delta[\eta(k)] g_\omega^*(r, -k; r'')$. Similarly, the complex conjugate of the Boltzmann equation for $g_\omega(r, -k; r'')$ is multiplied by $\delta[\eta(k)] g_\omega(r, k; r')$. The difference of the two equations obtained in this way is then integrated with respect to $r$ over the sample and with respect to $k$ over the whole $k$ space. Then use is made of some further consequences,

$$\eta(k) = \eta(-k), \quad v(k) = -v(-k), \tag{5.28}$$

of the invariance under time reversal. If again a certain boundary integral vanishes,

$$\sum_j \int d^3 k \, \delta[\eta_j(k)] \oint_{B_j} d^2 r \cdot v_j(k) g_{j\omega}^*(r, -k; r'') g_{j\omega}(r, k; r') = 0, \tag{5.29}$$

and if use is made of Eq. (4.31), the Hermiticity [Eq. (2.21)] follows apart from some changes in notation. For spherical Fermi surfaces, Eq. (5.29) may be written in the form

$$\sum_j N_j \oint d\Omega \oint_{B_j} d^2 r \cdot v_j g_\omega^*(r, -v_j; r'') g_\omega(r, v_j; r') = 0. \tag{5.30}$$

The $r$ integration in Eqs. (5.29) and (5.30) has to be extended over the complete surface of each separate metal. The lefthand side of Eqs. (5.29) and (5.30) will again vanish without the $r$ integration being performed. The physical meaning of these equations is that the normal component of the electric current density vanishes on external surfaces and is continuous on interfaces. This shall, however, not be shown before Section 18.

Combining Eq. (5.5) with Eqs. (5.25) and (5.28) we obtain

$$v(k) \sigma(k) = \int P(k, k') \delta[\eta(k) - \eta(k')] d^3 k'. \tag{5.31}$$

As a consequence of this equation, the micro-canonical ensemble (5.3) is, in the absence of a magnetic field, a stationary solution also to the time dependent Boltzmann equation (5.7) with scattering terms. For spherical Fermi surfaces, Eq. (5.31) may be written as

$$\sigma(v) = \oint \frac{d\sigma(v, v')}{d\Omega} \, d\Omega'. \tag{5.32}$$

No relation going beyond Eq. (5.13) is obtained for a differential scattering cross section depending only upon the scattering angle.

## 6. Boundary Conditions

The Boltzmann equation for $g_\omega(r, v; r')$, which was derived heuristically in the previous section, has to be supplemented by boundary conditions on external surfaces of the sample and on interfaces between different metals in electric contact. These boundary conditions shall be formulated in the present section. The way of arguing will again be heuristical. Only boundary conditions for specular reflexion could so far be derived on a microscopic quantum-mechanical basis [Section 7c]. But it is not clear how stochastic elements (non-specular reflexion) can be incorporated into a quantum-mechanical formulation.

If invariance under time reversal holds, in particular in the absence of a magnetic field, the function $g_\omega(r, v; r')$ may be interpreted as a distribution function of classical particles. This is the leading principle for the formulation of boundary conditions. The case of free electrons and of conduction electrons with spherical Fermi surface is particularly simple and imaginative. We shall mainly study this case and discuss the case of a general Fermi surface only very shortly afterwards.

### a) Surfaces

We start with the simpler problem of boundary conditions on an external surface [2]. For a specularly reflecting boundary, the classical particle picture leads to the boundary condition

$$g_\omega(r, v_{\text{out}}; r') = g_\omega(r, v_{\text{in}}; r'). \tag{6.1}$$

Here, $r'$ is an arbitrary point in the sample (the starting point) and $r$ is a point on the surface. $v_{\text{in}}$ is a velocity vector pointing from the interior towards the surface (incoming electrons), and $v_{\text{out}}$ is a velocity vector pointing from the surface into the interior (outgoing electrons). For specular reflexion, these two vectors with modulus equal to the Fermi velocity are related by

$$v_{\text{out}} = v_{\text{in}} - 2n(n \cdot v_{\text{in}}), \tag{6.2}$$

where $n$ is the normal unit vector on the surface in the point $r$. Eqs. (6.1) and (6.2) shall be assumed to be true also in the presence of a magnetic field when $g_\omega(r, v; r')$ can no longer be interpreted as a distribution function of electrons. This assumption essentially means that the reflexion process itself is not affected by the magnetic field. We might say that carriers of information are reflected like particles.

A specularly reflecting surface presents a particularly simple model. But physical surfaces do not seem to reflect electrons in a specular manner. A generalization of Eqs. (6.1) and (6.2) is definitely needed. In the general case, the velocity distribution of outgoing electrons will depend in a linear way upon the velocity distribution of incoming electrons. This dependence may formally be written as

$$g_\omega(r, v_{\text{out}}; r') = \int R(r; v_{\text{out}}, v_{\text{in}})\, g_\omega(r, v_{\text{in}}; r')\, d\Omega_{\text{in}} \qquad (6.3)$$

with integration over the semisphere of incoming directions. The reflexion kernel, $R(r; v_{\text{out}}, v_{\text{in}})$, characterizes the reflexion properties, for electrons with Fermi energy, of the surface in the point $r$. For specular reflexion as formulated by Eqs. (6.1) and (6.2) it is

$$R_{\text{sp}}(v_{\text{out}}, v_{\text{in}}) = \hat{\delta}(v_{\text{out}}, v_{\text{in}} - 2\, n(n \cdot v_{\text{in}})). \qquad (6.4)$$

The delta function of the directions of two vectors has been defined in Eq. (4.14). It shall again be assumed that Eq. (6.3) is, with unaltered reflexion kernel, also valid in the presence of a magnetic field. The position argument has been dropped in Eq. (6.4); this shall also be done in the further relations.

The reflexion kernel has to satisfy a number of general conditions which in principle are known already from Section 5. For the formulation of these conditions we again imagine the motion of classical electrons.

1. In this case, both $g_\omega(r, v_{\text{out}}; r')$ and $g_\omega(r, v_{\text{in}}; r')$ are non-negative quantities. We, therefore, have to require that it is

$$R(v_{\text{out}}, v_{\text{in}}) \geqq 0. \qquad (6.5)$$

This condition closely corresponds to Eq. (5.6).

2. Conservation of the number of particles requires that the surface normal component of the particle flux vanishes,

$$\oint n \cdot v\, g_\omega(r, v; r')\, d\Omega = 0. \qquad (6.6)$$

This equation is essentially equivalent to Eq. (5.24). Since Eq. (6.6) has to hold for an arbitrary distribution of the incoming directions of electrons, we postulate

$$\int n \cdot v_{\text{out}}\, R(v_{\text{out}}, v_{\text{in}})\, d\Omega_{\text{out}} = -\, n \cdot v_{\text{in}}. \qquad (6.7)$$

The analogous relation in the impurity scattering case is given by Eqs. (5.5), (5.10), and (5.13).

3. The postulate that the reflexion process be invariant under time reversal leads to

$$n \cdot v_{out} R(v_{out}, v_{in}) = - n \cdot v_{in} R(-v_{in}, -v_{out}) \tag{6.8}$$

which corresponds to Eqs. (5.25) and (5.26). The cosine terms, $n \cdot v_{out}$ and $n \cdot v_{in}$, are perhaps somewhat unexpected. Only with these additional factors the lefthand side of Eq. (5.30) vanishes without the $r$ integration being performed.

4. Inserting Eq. (6.8) into Eq. (6.7) we arrive at

$$\int R(v_{out}, v_{in}) \, d\Omega_{in} = 1 . \tag{6.9}$$

This equation, which is analogous to Eqs. (5.31) and (5.32), is certainly not independent of the three other conditions. It is, nevertheless, convenient to list it separately.

If a sample consists of a single metal there are only surfaces and no interfaces. The above general conditions together with the corresponding conditions on the impurity scattering parameters suffice to derive the general properties of the kernel, $K_\omega(r, r')$. In particular, Eq. (6.5) is required for proving that $g_\omega(r, v; r')$ is non-negative for $A(r) \equiv 0$. The sum rule (3.13) follows, for $A(r) \equiv 0$, from Eq. (6.7). The Hermiticity, Eq. (2.21), is derived with the help of Eq. (6.8). The micro-canonical ensemble (5.3) satisfies also the boundary conditions and, therefore, is a stationary solution to the time dependent Boltzmann equation, if Eq. (6.9) holds.

The kernel (6.4) of specular reflexion evidently satisfies all general conditions. If, by diffuse reflexion, we mean independence of the reflexion kernel upon $v_{out}$, the corresponding kernel follows uniquely [12],

$$R_{diff}(v_{out}, v_{in}) = \frac{n \cdot v_{in}}{\pi v} , \tag{6.10}$$

from the general conditions. Here $n$ is the normal unit vector pointing out of the sample. For many applications it suffices to work with a linear combination of the kernels of specular and of diffuse reflexion, i.e., with the one-parametric family

$$R(v_{out}, v_{in}) = (1 - p) R_{sp}(v_{out}, v_{in}) + p R_{diff}(v_{out}, v_{in}) . \tag{6.11}$$

The righthand side of Eq. (6.11) satisfies Eqs. (6.7)–(6.9). Also Eq. (6.5) is satisfied provided that the diffuseness, $p$, is bounded by

$$0 \le p \le 1 . \tag{6.12}$$

When the senior author [2] introduced this definition of $p$ he was, unfortunately, not aware of the opposite definition used by *Reuter* and *Sondheimer* [27] in their work on the anomalous skin effect. They called $p$ what we call $1 - p$. Since Eq. (6.11) has in the meantime been used in a number of papers on the influence of surface properties on critical fields, it seems unavoidable to stick to the present definition of $p$.

It is possible, though perhaps not of great practical importance, to generalize the relations to non-spherical Fermi surfaces. We denote by $k_{in}$ a momentum vector on the Fermi surface for which $v(k)$ as defined by Eq. (4.40) is an ingoing velocity. Similarly we define $k_{out}$. Then Eq. (6.3) is replaced by

$$g_\omega(r, k_{out}; r') = \int r(r; k_{out}, k_{in}) \, g_\omega(r, k_{in}; r') \, \delta[\eta(k_{in})] \, d^3 k_{in} \,. \tag{6.13}$$

The general conditions are formulated in the following way,

$$r(k_{out}, k_{in}) \gtrless 0 \,, \tag{6.14}$$

$$\int n \cdot v(k_{out}) \, r(k_{out}, k_{in}) \, \delta[\eta(k_{out})] \, d^3 k_{out} = - n \cdot v(k_{in}) \,, \tag{6.15}$$

$$n \cdot v(k_{out}) \, r(k_{out}, k_{in}) = - n \cdot v(k_{in}) \, r(-k_{in}, -k_{out}) \,, \tag{6.16}$$

$$\int r(k_{out}, k_{in}) \, \delta[\eta(k_{in})] \, d^3 k_{in} = 1 \,. \tag{6.17}$$

We shall not try to formulate analogues to specular and diffuse reflexion.

## b) Interfaces

If a sample consists of several metals in electric contact boundary conditions have also to be formulated on the interfaces between different metals [6]. As long as we are not interested in the microscopic details in the immediate neighbourhood of such interfaces, the Boltzmann equation for a homogeneous metal is valid everywhere inside each metal. Boundary conditions should be formulated on the interfaces themselves which are of a similar structure as Eq. (6.3). We shall only consider the case of spherical Fermi surfaces where each metal is characterized by any two of the three quantities: effective mass, $m^*$, Fermi velocity, $v$, and density of states, $N$; cf. Eq. (4.54). It is hoped that the final results (transition temperatures, critical fields) will be approximately true also for metals with Fermi surfaces of arbitrary shape.

We regard an interface between a metal number 1 and a metal number 2. There are electrons approaching the interface both from metal 1 and from metal 2 so that Eq. (6.3) is generalized to

$$\begin{aligned} g_\omega(r, v_{1\,out}; r') = &\int R_1(r; v_{1\,out}, v_{1\,in}) \, g_\omega(r, v_{1\,in}; r') \, d\Omega_{1\,in} \\ &+ \int T_1(r; v_{1\,out}, v_{2\,in}) \, g_\omega(r, v_{2\,in}; r') \, d\Omega_{2\,in} \,, \end{aligned} \tag{6.18}$$

where $r$ is a point on the interface.

The kernel $R_1(r; v_{1\,out}, v_{1\,in})$ characterizes the reflexion of electrons approaching the interface from metal 1. The kernel $T_1(r; v_{1\,out}, v_{2\,in})$ is characteristic for the transmission of electrons from metal 2 to metal 1. The two kernels shall be referred to as reflexion and transmission kernel, respectively. The modulus of all velocity vectors with subscript 1 is equal to the Fermi velocity, $v_1$, in metal 1. An analogous statement holds for all velocity vectors with subscript 2. Distribution functions containing a velocity vector $v_1$ are solutions to the Boltzmann equation for metal 1 and correspondingly for distribution functions containing a $v_2$. Eq. (6.18) is not yet complete. It has to be supplemented by a second relation,

$$g_\omega(r, v_{2\,\text{out}}; r') = \int R_2(r; v_{2\,\text{out}}, v_{2\,\text{in}}) \, g_\omega(r, v_{2\,\text{in}}; r') \, d\Omega_{2\,\text{in}}$$
$$+ \int T_2(r; v_{2\,\text{out}}, v_{1\,\text{in}}) \, g_\omega(r, v_{1\,\text{in}}; r') \, d\Omega_{1\,\text{in}} . \tag{6.19}$$

The reflexion and transmission kernels have to satisfy a number of general relations. Their conceptual basis is the same as that of Eqs. (6.5) and (6.7)–(6.9).

1. The non-negativeness of the distribution functions leads to

$$R_1(v_{1\,\text{out}}, v_{1\,\text{in}}) \geqq 0, \qquad T_1(v_{1\,\text{out}}, v_{2\,\text{in}}) \geqq 0 , \tag{6.20}$$

which corresponds to Eq. (6.5).

2. Conservation of particles requires the continuity of the normal component of the particle flux. We obtain

$$N_1 \{ n \cdot v_{1\,\text{in}} + \int n \cdot v_{1\,\text{out}} R_1(v_{1\,\text{out}}, v_{1\,\text{in}}) \, d\Omega_{1\,\text{out}} \}$$
$$= N_2 \int n \cdot v_{2\,\text{out}} T_2(v_{2\,\text{out}}, v_{1\,\text{in}}) \, d\Omega_{2\,\text{out}} \tag{6.21}$$

in close analogy to Eq. (6.7). It is remarkable that this relation contains the densities of states. If Eq. (6.21) and the corresponding equation with subscripts 1 and 2 interchanged are satisfied, the contributions to the lefthand side of Eq. (5.24) vanish in each point of the interface.

3. Invariance under time reversal is expressed by

$$n \cdot v_{1\,\text{out}} R_1(v_{1\,\text{out}}, v_{1\,\text{in}}) = - n \cdot v_{1\,\text{in}} R_1(-v_{1\,\text{in}}, -v_{1\,\text{out}}) \tag{6.22}$$

and

$$N_1 n_1 \cdot v_{1\,\text{out}} T_1(v_{1\,\text{out}}, v_{2\,\text{in}}) = N_2 n \cdot v_{2\,\text{in}} T_2(-v_{2\,\text{in}}, -v_{1\,\text{out}}) . \tag{6.23}$$

Eq. (6.22) is, apart from notational differences, identical with Eq. (6.8). If also the analogous relation with subscripts 1 and 2 is satisfied, the contributions to the lefthand side of Eq. (5.30) vanish in each point of the interface.

4. Inserting Eqs. (6.22) and (6.23) into Eq. (6.21), we obtain

$$\int R_1(v_{1\,\text{out}}, v_{1\,\text{in}}) \, d\Omega_{1\,\text{in}} + \int T_1(v_{1\,\text{out}}, v_{2\,\text{in}}) \, d\Omega_{2\,\text{in}} = 1 . \tag{6.24}$$

This relation is analogous to Eq. (6.9).

As already noticed above there are also relations corresponding to Eqs. (6.20), (6.21), (6.22), and (6.24) with subscripts 1 and 2 interchanged. It is evident that the general properties of the kernel, $K_\omega(r, r')$, may now be proved even if the sample consists of several metals in electric contact.

We shall not try to give certain parametrized standard forms for the reflexion and transmission kernels in analogy to Eq. (6.11). Though this could easily be done for the reflexion kernels [the righthand side of Eq. (6.11) would have to be multiplied by a factor between zero and one; cf. Eqs. (15.83) and (15.84)], it would be a difficult task for the transmission kernels. We shall only try to discuss the Eqs. (6.18)–(6.24) on a general basis.

If there is no electric contact between the two metals, the transmission kernels vanish identically. In this case the above equations reduce to the boundary conditions, Eqs. (6.3) ff., on external surfaces. On the other hand, ideal electric contact does not mean that both reflexion kernels vanish identically. Vanishing reflexion kernels are possible only if

$$N_1 v_1 = N_2 v_2 \qquad (6.25)$$

is satisfied. For the proof it is only necessary to integrate Eq. (6.21), with $R_1(v_{1\,\text{out}}, v_{1\,\text{in}})$ put equal to zero, over all directions of $v_{1\,\text{in}}$ and to make use of Eq. (6.24) with the subscripts 1 and 2 interchanged and with $R_2(v_{2\,\text{out}}, v_{2\,\text{in}})$ put equal to zero.

A physical insight into Eq. (6.25) is easily obtained. We regard an interface without stochastic elements. The electronic orbits will undergo a refraction in such a way that the tangential component of the momentum is continuous. With the angles, $\theta_1$ and $\theta_2$, measured with respect to the interface normal, it is

$$\frac{\sin \theta_1}{\sin \theta_2} = \frac{m_2^* v_2}{m_1^* v_1} = \sqrt{\frac{N_2 v_2}{N_1 v_1}} . \qquad (6.26)$$

Only in the special case given by Eq. (6.25), all orbits will pass unrefracted through the interface. In all other cases there are always orbits, corresponding to total reflexion in the optical analogue, which cannot penetrate into the metal with smaller value of $Nv$.

## 7. Proofs for Free Electrons

The general formulation developed in the previous three sections was based on somewhat vague intuitive arguments: The kernel $K_\omega(r, r')$ is, in the time reversal invariant case, the Laplace transform of the slowly varying part of a quantum mechanical correlation function [Eq. (4.4)]. This slowly varying part was tentatively interpreted as the corresponding classical correlation function [Eq. (4.6)], which may be obtained from a distribution function in phase space [Eq. (4.29)]. Impurity scattering was introduced in a statistical manner by postulating the Boltzmann equation (5.11) for the distribution function. In the same spirit, boundary conditions on surfaces [Eq. (6.3)] and interfaces [Eq. (6.18)] were put forward. Finally, the formulation was extended to the case of an external magnetic field by taking into account the gauge properties [Eq. (3.2)] of $K_\omega(r, r')$ in the simplest possible way [Eq. (5.16)].

This chain of arguments has turned out useful since it led to a rather simple formulation which appeals to intuition. However, proofs, which should be based upon the microscopic quantum-mechanical theory, have so far not been supplied in this review article. This task shall now be performed. In this section the model of free electrons will be studied. A justification will be given [3] for the introduction of a distribution function and for the Boltzmann equation obeyed by this distribution function. It will also turn out possible to prove the boundary condition on external specularly reflecting surfaces. A somewhat less conclusive analysis of conduction electrons in periodic lattices will be given in the next section. Finally, in Section 9 the Boltzmann equation will be derived for the case of paramagnetic impurities.

### a) No Impurities Present

The proof of the Boltzmann equation proceeds in several steps. We have already seen in Section 4c that the classical treatment leads to the correct result for $K_\omega(r, r')$ in the absence of impurities. Also the distribution function has already been written down in Eq. (4.38). It should, however, be remembered that the coincidence of classical and quantum mechanical results was only obtained under the condition (4.9) which may also be written in the form

$$k_F \xi_\omega \gg 1.  \tag{7.1}$$

Here, $k_F$ is the Fermi momentum [Eq. (3.18)] and $\xi_\omega$ is a length defined by

$$\xi_\omega := \frac{v}{2|\omega|}.  \tag{7.2}$$

The inequality (7.1) is satisfied even for the smallest $\xi_\omega$ entering into the calculations (i.e., for $|\omega| = \omega_D$), since the Fermi velocity is big as compared to the velocity of sound.

That the distribution function (4.38) satisfies the Boltzmann equation

$$\left(2|\omega| + v \cdot \frac{\partial}{\partial r}\right) g_\omega(r, v; r') = \frac{1}{(2\pi)^2} \delta(r - r')  \tag{7.3}$$

is most easily shown in an indirect manner. The time dependent distribution function $f(r, k; r'; t)$ as defined by Eqs. (4.30) and (4.35) evidently satisfies the time dependent Boltzmann (Liouville) equation (5.1) without $k$ derivative, and the initial condition (4.33). If $g_\omega(r, v; r')$ is introduced through Eq. (4.31) with the result (4.36), Eq. (7.3) follows immediately.

A direct proof is obtained by first showing that it is

$$\left(\boldsymbol{v} \cdot \frac{\partial}{\partial \boldsymbol{r}}\right) \frac{1}{|\boldsymbol{r}-\boldsymbol{r'}|^2} \hat{\delta}(\boldsymbol{r}-\boldsymbol{r'}, \boldsymbol{v}) = 0. \tag{7.4}$$

(Choose the $z$ direction parallel to the vector $\boldsymbol{v}$; then it is

$$\boldsymbol{v} \cdot \frac{\partial}{\partial \boldsymbol{r}} = v \frac{\partial}{\partial z} \tag{7.5}$$

and

$$\hat{\delta}(\boldsymbol{v}, \boldsymbol{r}-\boldsymbol{r'}) = |\boldsymbol{r}-\boldsymbol{r'}|^2 \, \delta(x-x') \, \delta(y-y').) \tag{7.6}$$

It is further easily seen that it is

$$\left(2|\omega| + \boldsymbol{v} \cdot \frac{\partial}{\partial \boldsymbol{r}}\right) \exp\left[-\frac{2|\omega| \, |\boldsymbol{r}-\boldsymbol{r'}|}{v}\right] = 0, \tag{7.7}$$

provided that the vectors $\boldsymbol{v}$ and $\boldsymbol{r}-\boldsymbol{r'}$ are parallel to each other. The inhomogeneity on the righthand side of Eq. (7.3) is obtained from the discontinuity of $g_\omega(\boldsymbol{r}, \boldsymbol{v}; \boldsymbol{r'})$ [Eq. (4.38)] at $|\boldsymbol{r}-\boldsymbol{r'}|=0$ with the help of Eq. (7.6) and of

$$\delta(x-x') \, \delta(y-y') \, \delta(z-z') = \delta(\boldsymbol{r}-\boldsymbol{r'}). \tag{7.8}$$

In the presence of a magnetic field, the Green's function $G_\omega^{(0)}(\boldsymbol{r}, \boldsymbol{r'})$ satisfies Eq. (2.25). This equation may be solved in a quasi-classical approximation [3]. We regard the classical motion of an electron from the position $\boldsymbol{r'}$ to the position $\boldsymbol{r}$. The energy of the particle will be fixed later. With the help of

$$\boldsymbol{p}(s) = m\boldsymbol{v}(s) - e\boldsymbol{A}(s), \tag{7.9}$$

the canonical momentum in an arbitrary point, $s$, on the orbit, we introduce the action integral,

$$S(\boldsymbol{r}, \boldsymbol{r'}) = \int_{\boldsymbol{r'}}^{\boldsymbol{r}} \boldsymbol{p}(s) \cdot \mathrm{d}s, \tag{7.10}$$

with integration along the classical orbit. As a consequence of the equation of motion (or of the principle of least action), it is

$$\frac{\partial S(\boldsymbol{r}, \boldsymbol{r'})}{\partial \boldsymbol{r}} = \boldsymbol{p}(\boldsymbol{r}), \qquad \frac{\partial S(\boldsymbol{r}, \boldsymbol{r'})}{\partial \boldsymbol{r'}} = -\boldsymbol{p}(\boldsymbol{r'}), \tag{7.11}$$

where the differentiations are done with the energy of the electron and the other end of the orbit kept fixed.

We now make the ansatz

$$G_\omega^{(0)}(\boldsymbol{r}, \boldsymbol{r'}) = f_\omega(\boldsymbol{r}) \exp[i S(\boldsymbol{r}, \boldsymbol{r'})], \tag{7.12}$$

where the function $f_\omega(r)$, as a consequence of Eqs. (2.25) and (7.12), has
to satisfy

$$\frac{1}{2m} \left( \frac{\partial}{\partial r} \right)^2 f_\omega(r) + i \left[ v(r) \cdot \frac{\partial f_\omega(r)}{\partial r} + \frac{1}{2} \frac{\partial v(r)}{\partial r} f_\omega(r) + \omega f_\omega(r) \right]$$
$$+ \left( \mu - \frac{m v^2}{2} \right) f_\omega(r) = \delta(r - r'). \tag{7.13}$$

In order to solve this differential equation, we firstly require

$$\frac{m v^2}{2} = \mu, \tag{7.14}$$

so that the classical motion is executed with Fermi energy. Secondly,
we postulate

$$v(r) \cdot \frac{\partial f_\omega(r)}{\partial r} + \frac{1}{2} \frac{\partial v(r)}{\partial r} f_\omega(r) + \omega f_\omega(r) = 0. \tag{7.15}$$

If the orbit may be approximated by a straight line, i.e., if it is

$$v(r) = v \frac{r - r'}{|r - r'|}, \tag{7.16}$$

it follows that it is

$$\frac{\partial v(r)}{\partial r} = \frac{2v}{|r - r'|}. \tag{7.17}$$

Eq. (7.15) may now be integrated along the classical orbit with the result

$$f_\omega(r) = \frac{\text{const}}{|r - r'|} \exp \left[ -\frac{\omega |r - r'|}{v} \right]. \tag{7.18}$$

Thirdly, the constant is determined in such a way that

$$\frac{1}{2m} \left( \frac{\partial}{\partial r} \right)^2 f_\omega(r) = \delta(r - r') \tag{7.19}$$

is approximately satisfied. Differentiating only the denominator (i.e. the
part becoming singular for $r \to r'$) on the righthand side of Eq. (7.18),
we finally obtain

$$f_\omega(r) = -\frac{m}{2\pi |r - r'|} \exp \left[ -\frac{\omega |r - r'|}{v} \right]. \tag{7.20}$$

Collecting all factors we derive the following result,

$$G_\omega^{(0)}(r, r') = -\frac{m}{2\pi |r - r'|} \exp \left[ imv|r - r'| - ie \int_{r'}^{r} A(s) \cdot ds - \frac{\omega}{v} |r - r'| \right], \tag{7.21}$$

of the quasi-classical approximation. The integral in the exponential is extended along the straight line connecting the position $r'$ with the position $r$.

This approximate result shows the correct behaviour under gauge transformations [Eq. (3.2)]. For vanishing vector potential it reduces to Eq. (4.10). Since $G_\omega^{(0)}(r, r')$ contains an exponential damping factor with characteristic length $2\xi_\omega$ [Eq. (7.2)], the approximation of replacing the curved orbit by a straight line is justified only if $2\xi_\omega$ is small as compared to the radius of curvature of the electronic orbit,

$$2\xi_\omega \ll \frac{mv}{eH}. \tag{7.22}$$

Eq. (7.22), or

$$|\omega| \gg \frac{eH}{m}, \tag{7.23}$$

may also be interpreted in a somewhat different manner. It means that, in the energy denominator of Eq. (3.9), the discrete nature of the Landau levels may be disregarded. The Landau wave functions are, however, in an approximative way contained in the righthand side of Eq. (7.21).

The righthand side of Eq. (7.21) is bounded for $|r - r'| \to \infty$ only if $\omega$ is a non-negative quantity. In the case $\omega < 0$ the righthand side of Eq. (7.12) is replaced by

$$G_\omega^{(0)}(r, r') = \tilde{f}_\omega(r) \exp \left[ -iS(r', r) \right]. \tag{7.24}$$

The analogous procedure as above then leads to

$$G_\omega^{(0)}(r, r') = -\frac{m}{2\pi|r - r'|} \exp\left[ -imv|r - r'| + ie \int_r^{r'} A(s) \cdot ds + \frac{\omega|r - r'|}{v} \right]. \tag{7.25}$$

Eqs. (7.21) and (7.25) together obviously satisfy Eq. (2.21).

As long as the curvature of the orbit in the magnetic field is neglected, both factors on the righthand side of Eq. (2.16) refer to the same straight orbit so that $K_\omega(r, r')$ may again be expressed by a (no longer real) distribution function [Eq. (4.39)]. The comparison of Eq. (7.21) with Eq. (4.10) and the consideration of Eq. (4.38) lead to

$$g_\omega(r, v; r') = \frac{1}{(2\pi|r - r'|)^2 v} \exp\left[ -\frac{2|\omega| \, |r - r'|}{v} - 2ie \int_{r'}^r A(s) \cdot ds \right] \delta(r - r', v), \tag{7.26}$$

where the charge $-2e$ in the exponential reflects the fact that the carriers of charge in the superconductor are electron pairs. The distribution function (7.26) obviously satisfies

$$(2|\omega| + v \cdot \tilde{\partial}) \, g_\omega(r, v; r') = \frac{1}{(2\pi)^2} \delta(r - r'). \tag{7.27}$$

The gauge invariant gradient, $\tilde{\partial}$, was defined in Eq. (5.16). If the magnetic curvature of the orbit may not be neglected, no simple result is obtained.

We do, in particular, not arrive at a Boltzmann equation with a Lorentz force term.

The quasi-classical approximation is, even in the absence of a magnetic field, not restricted to a position independent Hamiltonian. In the more general case [3], a Boltzmann equation is obtained which contains a force term. This generalization is, however, of no practical interest for the superconductivity problem.

### b) Impurity Averaged Green's Function

Now impurity atoms are introduced. They shall be simply treated as scattering potentials. We assume that the magnetic field is again absent. The matrix element of the one-electron Hamiltonian which has to be inserted into Eq. (1.23) is given by

$$\langle r|h| r''\rangle = \left( - \frac{1}{2m} \left( \frac{\partial}{\partial r} \right)^2 + \sum_j v(r - r_j) \right) \delta(r - r''). \qquad (7.28)$$

Here, $v(r - r_j)$ is the scattering potential of an impurity atom situated at the position $r_j$. There would certainly be no point in first specifying all positions $r_j$ and then calculating the quantities of interest for this specified distribution. Meaningful results will only be obtained if an average is performed with respect to a statistical distribution of impurity atoms. It would, however, be wrong to average directly the righthand side of Eq. (7.28). The solutions have first to be worked out assuming some specified distribution of the impurity positions. The averaging has then to be done at the very end of the calculation.

Before the averages are formed, the Green's function, $G_\omega^{(0)}(r, r')$, the kernel, $K_\omega(r, r')$, and the pair potential, $\Delta(r)$, depend explicitly upon all positions $r_j$ though this dependence shall not be exhibited on the symbols for Green's function, kernel, and pair potential. If the average with respect to a probability distribution of impurity atoms is denoted by angular brackets, the gap equation (2.15) reads

$$\langle \Delta(r)\rangle = g(r) \int T \sum_\omega{}' \langle G_\omega^{(0)}(r, r') G_\omega^{(0)*}(r', r) \Delta(r')\rangle \, \mathrm{d}^3 r'. \qquad (7.29)$$

Fortunately enough, it turns out [28] that this equation may be replaced by

$$\langle \Delta(r)\rangle = g(r) \int T \sum_\omega{}' \langle G_\omega^{(0)}(r, r') G_\omega^{(0)*}(r', r)\rangle \langle \Delta(r')\rangle \, \mathrm{d}^3 r', \qquad (7.30)$$

as long as

$$k_\mathrm{F} l \gg 1 \qquad (7.31)$$

holds. Here, $l$ is the electronic mean free path due to impurity scattering. The impurity averaged pair potential shall from now on again be denoted by $\Delta(r)$ without angular brackets. Our task will be to calculate the impurity averaged kernel,

$$K_\omega(r, r') = \langle G_\omega^{(0)}(r, r')\, G_\omega^{(0)*}(r', r) \rangle\,. \qquad (7.32)$$

We shall first calculate the impurity averaged Green's function, $\langle G_\omega^{(0)}(r, r')\rangle$, though it would certainly be wrong to replace the righthand side of Eq. (7.32) by the product of two impurity averaged Green's functions. The Green's function of normal electrons with impurities absent shall now be denoted by $G_\omega^{(0\,0)}(r, r')$. In a homogeneous metal it is given by the righthand side of Eq. (4.10). If the potential produced by all impurity atoms is called

$$W(r) = \sum_j v(r - r_j)\,, \qquad (7.33)$$

the Green's function, $G_\omega^{(0)}(r, r')$, for a specified distribution of the positions $r_j$ satisfies the integral equation

$$G_\omega^{(0)}(r, r') = G_\omega^{(0\,0)}(r, r') + \int G_\omega^{(0\,0)}(r, r'')\, W(r'')\, G_\omega^{(0)}(r'', r')\, d^3 r''. \qquad (7.34)$$

This equation follows, with the appropriate changes in notation, directly from Eq. (1.36). We imagine this integral equation to be solved in an iterative manner and the result to be averaged with respect to a probability distribution of the positions, $r_j$.

If the action of an individual scattering potential is, as usual [29, 30], taken into account up to second order, a few low order contributions to $\langle G_\omega^{(0)}(r, r')\rangle$ are, in graphical notation, given by

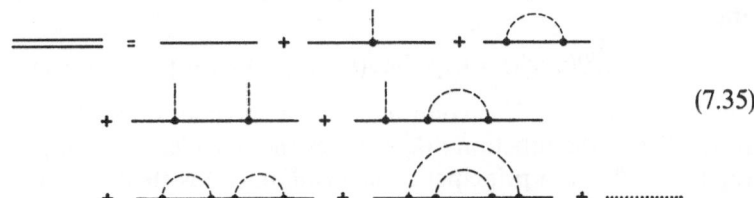

$$(7.35)$$

The double line on the lefthand side denotes $\langle G_\omega^{(0)}(r, r')\rangle$ whereas a single solid line represents $G_\omega^{(0\,0)}(r, r')$. The dotted lines are impurity lines. If only one end is attached to the solid line, the scattering by a certain impurity atom occurs only once. If both ends are connected to a solid line, the same impurity atom scatters the electron twice. Contributions from graphs with intersecting impurity lines are negligeable [28] as long as Eq. (7.31) holds.

All contributions to $\langle G_\omega^{(0)}(r, r')\rangle$ which shall be taken into account can obviously also be obtained as the iterative solution to

$$\tag{7.36}$$

or

$$\langle G_\omega^{(0)}(r, r')\rangle = G_\omega^{(00)}(r, r') + \int G_\omega^{(00)}(r, s)\, n(s)\, [\bar{v}\delta(s - s')$$
$$+ w(s - s')\langle G_\omega^{(0)}(s, s')\rangle]\, \langle G_\omega^{(0)}(s', r')\rangle\, d^3s\, d^3s' . \tag{7.37}$$

Here, $n(s)$ is the (possibly on a macroscopic scale position dependent) concentration (number per volume) of impurity atoms. Further it is

$$\bar{v} = \int v(r)\, d^3r, \quad w(r - r') = \int v(r - t)\, v(r' - t)\, d^3t , \tag{7.38}$$

where the space integrations are consequences of the averaging procedure. Since $G_\omega^{(00)}(r, r')$ satisfies Eq. (4.7), we obtain

$$\left(i\omega + \mu + \frac{1}{2m}\left(\frac{\partial}{\partial r}\right)^2\right)\langle G_\omega^{(0)}(r, r')\rangle = \delta(r - r') + n(r)\,\bar{v}\langle G_\omega^{(0)}(r, r')\rangle$$
$$+ n(r)\int w(r - s')\langle G_\omega^{(0)}(r, s')\rangle\, \langle G_\omega^{(0)}(s', r')\rangle\, d^3s' . \tag{7.39}$$

The only remaining task is the evaluation of the integral in this equation.

Because of the short range of the scattering potential, the integrand is appreciable only if the points $r$ and $s'$ are close together. We tentatively put

$$\langle G_\omega^{(0)}(r, s')\rangle = -\frac{1}{(2\pi)^3}\int d^3k\, \exp[ik\cdot(r - s')]\left\{\frac{P}{\eta(k)} + i\pi\,\delta[\eta(k)]\, \mathrm{sgn}\,\omega\right\}$$
$$\tag{7.40}$$

and

$$\langle G_\omega^{(0)}(s', r')\rangle = \exp[imv(r)\cdot(s' - r)\,\mathrm{sgn}\,\omega]\,\langle G_\omega^{(0)}(r, r')\rangle . \tag{7.41}$$

These equations will be confirmed below in small print after Eq. (7.55). In Eq. (7.40), the function $\eta(k)$ denotes the one-electron energy given in Eq. (3.17). $P$ is the principal value symbol. $v(r)$ is the velocity vector of a classical electron in the point $r$ which has moved on a straight line from $r'$ to $r$ with Fermi velocity.

We introduce the Fourier transform of the scattering potential by

$$v(r) = \frac{1}{(2\pi)^3}\int v(k)\, \exp(ik\cdot r)\, d^3k . \tag{7.42}$$

Then it is

$$\int w(k)\, \exp(-ik\cdot r)\, d^3r = |v(k)|^2 . \tag{7.43}$$

We note for future use that the differential scattering cross section is, in Born approximation, given by

$$\frac{d\sigma(v, v')}{d\Omega} = \left(\frac{m}{2\pi}\right)^2 |v(k - k')|^2 . \tag{7.44}$$

Here, $v$ and $k$, $v'$ and $k'$, are related to each other by Eq. (4.34). The modulus of the velocity vectors, $v$ and $v'$, equals the Fermi velocity, $v$, whereas the modulus of the momentum vectors, $k$ and $k'$, equals the Fermi momentum, $k_F$. The total scattering cross section in Born approximation is, as a consequence of Eq. (5.10), given by

$$\sigma(v) = \oint \frac{d\sigma(v', v)}{d\Omega'} d\Omega' = \frac{1}{(2\pi)^3 v} \int |v(k - k')|^2 \delta[\eta(k')] d^3 k'. \tag{7.45}$$

It follows immediately from Eq. (7.44) that the first Eq. (5.26) is satisfied in Born approximation. Since the scattering potential, $v(r)$, is a real quantity it further follows from Eq. (7.42) that it is

$$v(k) = v^*(-k). \tag{7.46}$$

Eq. (7.44) then leads to

$$\frac{d\sigma(v', v)}{d\Omega} = \frac{d\sigma(-v', -v)}{d\Omega}. \tag{7.47}$$

The second Eq. (5.26) is finally obtained with the help of Eq. (5.10).

With the help of Eqs. (7.40), (7.41), and (7.43) the integral in Eq. (7.39) may be rewritten in the form

$$\int d^3 s' w(r - s') \langle G_\omega^{(0)}(r, s') \rangle \langle G_\omega^{(0)}(s', r') \rangle$$

$$= -\frac{1}{(2\pi)^3} \int d^3 k |v(mv(r)\mathrm{sgn}\omega - k)|^2 \left\{ \frac{P}{\eta(k)} + i\pi\delta[\eta(k)]\mathrm{sgn}\omega \right\} \langle G_\omega^{(0)}(r, r') \rangle . \tag{7.48}$$

Both the principal value integral and the term $n\bar{v}$ may, in Eq. (7.39), be discarded as shall be justified in some detail further below in small print. The delta function integral may be expressed by the scattering cross section in Born approximation, Eq. (7.45). In this way we obtain

$$\left(i\omega + \frac{i}{2} n(r) v\sigma[v(r)] \,\mathrm{sgn}\,\omega + \mu + \frac{1}{2m} \left(\frac{\partial}{\partial r}\right)^2\right) \langle G_\omega^{(0)}(r, r') \rangle = \delta(r - r') \tag{7.49}$$

from Eq. (7.39). A comparison with Eqs. (4.7) and (4.10) shows that the solution is given by

$$\langle G_\omega^{(0)}(r, r') \rangle = -\frac{m}{2\pi|r - r'|} \exp\left[\left(imv\,\mathrm{sgn}\,\omega - \frac{n\sigma[v(r)]}{2} - \frac{|\omega|}{v}\right)|r - r'|\right] \tag{7.50}$$

for position independent concentration of impurities, and by

$$\langle G_\omega^{(0)}(r, r')\rangle$$

$$= - \frac{m}{2\pi|r - r'|} \exp\left[\left(imv\,\text{sgn}\,\omega - \frac{|\omega|}{v}\right)|r - r'| - \int_{r'}^{r} \frac{n(s)\,\sigma[v(s)]}{2}\,|ds|\right]$$

$$(7.51)$$

for variable $n(s)$ with integration along the classical (straight) electronic orbit. Eqs. (7.50) and (7.51) become somewhat simpler if the scattering cross section, $\sigma$, does not depend upon the direction of incoming electrons. In this case, the mean free path as occurring in Eq. (7.31) is given by

$$l := \frac{1}{n\sigma}. \tag{7.52}$$

It follows from Eq. (7.50) that the decay length of $\langle G_\omega^{(0)}(r, r')\rangle$ is no longer given by $2\xi_\omega$ [Eq. (7.2)] but by $2\zeta_\omega$ with

$$\frac{1}{\zeta_\omega} := \frac{1}{\xi_\omega} + \frac{1}{l}. \tag{7.53}$$

$\zeta_\omega$ approaches $l$ for sufficiently short mean free path. The condition for the validity of Eqs. (7.50) and (7.51) is

$$k_F \zeta_\omega \gg 1 \tag{7.54}$$

which replaces Eq. (7.1). Eq. (7.31) is a consequence of this condition. In the presence of a magnetic field, the quasi-classical approximation with neglection of the curvature of the orbit will be valid for

$$2\zeta_\omega \ll \frac{mv}{eH}; \tag{7.55}$$

cf. Eq. (7.22).

It is now easy to confirm Eqs. (7.40) and (7.41) on the basis of Eqs. (7.50) and (7.51). Eq. (7.40) is a Fourier integral representation in which approximations have been made which are permitted for small distance between the arguments of the Green's function provided that Eq. (7.54) holds. Eq. (7.41) is obtained if the approximation

$$v|s' - r'| = v(s') \cdot (s' - r') \approx v(r) \cdot (s' - r) = v(r) \cdot (s' - r) + v|r - r'| \tag{7.56}$$

is made. Eq. (7.56) is not valid for $r$ and $r'$ close together. But it is hoped that the error made in this way does not affect much the final results (7.50) and (7.51).

A comparison of Eqs. (7.50) and (4.10) and a recourse to Eq. (4.39) show that there exists a distribution function, $g_\omega^{(\sigma)}(r, v; r')$, so that it is

$$\langle G_\omega^{(0)}(r, r')\rangle \langle G_\omega^{(0)*}(r', r)\rangle = m^2 v \oint g_\omega^{(\sigma)}(r, v; r')\,d\Omega. \tag{7.57}$$

$g_\omega^{(\sigma)}(r, v; r')$ satisfies

$$\left(2|\omega| + v \cdot \frac{\partial}{\partial r} + n(r) \, v \, \sigma(v)\right) g_\omega^{(\sigma)}(r, v; r') = \frac{1}{(2\pi)^2} \, \delta(r - r') \quad (7.58)$$

even if $n(r)$ is not constant in space. $g_\omega^{(\sigma)}(r, v; r')$ obviously is the distribution function of electrons starting from $r'$ and reaching $r$ without having suffered a single impurity scattering.

The potential like terms ($n\bar{v}$ and principal value integral) in Eq. (7.39) have been discarded in the course of the derivation of Eq. (7.49). If these terms are position independent (constant concentration of impurities) and also do not depend upon the direction of the electron (spherically symmetric scattering potential) they may be incorporated into a "renormalization of the chemical potential" [30]. The situation is more complicated if these terms are not position and direction independent. In this case, they give rise to a force acting upon the electrons. This force is counteracted by another force due to an electric field which is created by a violation of local electric neutrality. Since these forces are not expected to be of relevance for the phenomenon of superconductivity, they are simply disregarded.

### c) Impurity Averaged Kernel

The Eqs. (7.57) and (7.58) do not solve our problem completely since we have to calculate the righthand side of Eq. (7.32) which obviously is not identical with the lefthand side of Eq. (7.57). We introduce the function

$$K_\omega(r, r', r'') = \langle G_\omega^{(0)}(r, r') \, G_\omega^{(0)*}(r', r'') \rangle. \quad (7.59)$$

It reduces to $K_\omega(r, r')$ for $r'' = r$. If the function $K_\omega(r, r', r'')$ is graphically represented by a box, the analogue to Eq. (7.35) is given by

$$(7.60)$$

We have already discarded graphs in which only one end of an impurity line is fixed to a Green's function line or in which impurity lines cross. If both ends of an impurity line are connected to the same Green's function, a contribution to $\langle G_\omega^{(0)}(r, r') \rangle$ is obtained.

More interesting are graphs in which both ends of an impurity line are attached to different Green's function lines. These additional terms are sometimes called vertex contributions. Evidently, the righthand side of Eq. (7.60) is obtained as the iterative solution to

$$(7.61)$$

or

$$
K_\omega(r, r', r'') = \langle G_\omega^{(0)}(r, r')\rangle \langle G_\omega^{(0)*}(r', r'')\rangle
$$
$$
+ \int d^3s\, d^3s'' \langle G_\omega^{(0)}(r, s)\rangle\, n(s)\, w(s - s'')\, K_\omega(s, r', s'') \langle G_\omega^{(0)*}(s'', r'')\rangle . \qquad (7.62)
$$

These equations correspond to Eqs. (7.36) and (7.37), respectively.

The main contributions to the integral in Eq. (7.62) come again from neighbouring points $s$ and $s''$ because of the small range of the scattering potential. Therefore, also neighbouring points $r$ and $r''$ are of main interest for solving the integral equation (7.62). By a slight generalization of Eq. (7.41) we may put

$$
\langle G_\omega^{(0)*}(s'', r'')\rangle = \exp\left[im\,v\,\frac{r - s}{|r - s|}\cdot(r - r'' + s'' - s)\right] \langle G_\omega^{(0)*}(s, r)\rangle . \qquad (7.63)
$$

Here and in the further equations, we have put $\omega > 0$. It then follows from Eq. (7.57) that it is

$$
\langle G_\omega^{(0)}(r, s)\rangle \langle G_\omega^{(0)*}(s'', r'')\rangle
$$
$$
= m^2 v \oint \exp[im\,v\cdot(r - r'' + s'' - s)]\, g_\omega^{(\sigma)}(r, v; s)\, d\Omega . \qquad (7.64)
$$

In this way, the righthand side of Eq. (7.62) may be expressed by a distribution function,

$$
K_\omega(r, r', r'') = m^2 v \oint \exp[im\,v\cdot(r - r'')]\, g_\omega(r, v; r')\, d\Omega , \qquad (7.65)
$$

which is implicitly given by

$$
g_\omega(r, v; r') = g_\omega^{(\sigma)}(r, v; r')
$$
$$
+ \int d^3s\, d^3s'' g_\omega^{(\sigma)}(r, v; s)\, n(s) \exp[im\,v\cdot(s'' - s)]\, w(s - s'')\, K_\omega(s, r', s'') . \qquad (7.66)
$$

Putting $r''$ equal to $r$ in Eq. (7.65), we see that $g_\omega(r, v; r')$ actually is the distribution function we are looking for.

Inserting Eq. (7.65) again into the righthand side of Eq. (7.66) and evaluating the integral with the help of Eq. (7.44) we obtain the integral equation

$$
g_\omega(r, v; r') = g_\omega^{(\sigma)}(r, v; r')
$$
$$
+ (2\pi)^2 \int d^3s\, g_\omega^{(\sigma)}(r, v; s)\, n(s)\, v \oint d\Omega'\, \frac{d\sigma(v, v')}{d\Omega}\, g_\omega(s, v'; r') . \qquad (7.67)
$$

The angular integration is extended over all directions of $v'$. Applying Eq. (7.58), we finally obtain the Boltzmann equation (5.11) which, for an angle dependent differential scattering cross section, goes over into Eq. (5.14). It might be noticed that the second scattering term on the lefthand side (scattering from $v'$ to $v$) in these equations follows entirely from the vertex contributions.

It is not difficult to include a magnetic field in the considerations. In the sense of the quasi-classical approximation, $G_\omega^{(0\,0)}(r, r')$ now contains a phase factor as in Eq. (7.21). Most other considerations remain the same apart from some notational complications. It has to be kept in mind that the relation between canonical momentum and velocity is now given by Eq. (7.9). In this way, finally

$$(2|\omega| + v \cdot \tilde{\partial} + nv\sigma(v))\, g_\omega(r, v; r')$$
$$- nv\oint \frac{d\sigma(v, v')}{d\Omega}\, g_\omega(r, v'; r')\, d\Omega' = \frac{1}{(2\pi)^2}\, \delta(r - r') \tag{7.68}$$

is obtained. For a differential scattering cross section depending only upon the scattering angle, it reduces to

$$(2|\omega| + v \cdot \tilde{\partial} + nv\sigma)\, g_\omega(r, v; r')$$
$$- nv\oint \frac{d\sigma(\theta)}{d\Omega}\, g_\omega(r, v'; r')\, d\Omega' = \frac{1}{(2\pi)^2}\, \delta(r - r'). \tag{7.69}$$

The symbol $\tilde{\partial}$ has been explained in Eq. (5.16).

### d) Boundary Conditions for Specular Reflexion

A general proof of the boundary conditions cannot be given since it is not clear how stochastic elements in the surface or interface scattering can be incorporated correctly into the microscopic quantum mechanical theory. We shall only try to prove the boundary condition (6.1), (6.2) for specular reflexion. In order to simplify the arguments we assume a plane boundary. Further, only a metal without impurities is regarded.

If the boundary is represented by an infinitely high potential, Eq. (2.25) has to be solved with the boundary condition

$$G_\omega^{(0)}(r, r') = 0. \tag{7.70}$$

Here, $r'$ is an arbitrary point in the sample and $r$ is a point on the surface. We want to produce a solution to Eqs. (2.25) and (7.70) with the help of the quasi-classical approximation.

If only a single plane boundary is taken into account, there are two classical orbits leading from the point $r'$ to the point $r$, one direct straight orbit and one orbit which is specularly reflected by the surface. The solution is, in quasi-classical approximation, given by

$$G_\omega^{(0)}(r, r') = G_{\omega\,\mathrm{dir}}^{(0)}(r, r') - G_{\omega\,\mathrm{refl}}^{(0)}(r, r').\qquad (7.71)$$

The first term on the righthand side corresponds to the direct orbit and is given by Eq. (7.21). The delta function on the righthand side of Eq. (2.25) is produced solely by this term. In $G_{\omega\,\mathrm{refl}}^{(0)}(r, r')$, which corresponds to the reflected orbit, the quantity $|r - r'|$ is replaced by the length of the actual orbit. The integral containing the vector potential is also taken along the actual reflected orbit. The minus sign is inserted into Eq. (7.71) in order to satisfy the boundary condition (7.70): direct and reflected orbit become identical for $r$ on the boundary. The second term on the righthand side of Eq. (7.71) would become more complicated if the surface were not plane. The quantity $\partial v(r)/\partial r$ in Eq. (7.13) would have to be calculated for a bundle of reflected orbits which all start from the point $r'$.

Since the individual Green's function on the righthand side of Eq. (7.71) is a rapidly oscillating function and since $K_\omega(r, r')$ is in Eq. (2.15) multiplied by a sufficiently slowly varying function of $r'$ before an integration with respect to $r'$ is carried out, we may approximately put

$$
\begin{aligned}
K_\omega(r, r') = G_\omega(r, r')\, G_\omega^*(r', r) &\approx G_{\omega\,\mathrm{dir}}^{(0)}(r, r')\, G_{\omega\,\mathrm{dir}}^{(0)*}(r', r)\\
&+ G_{\omega\,\mathrm{refl}}^{(0)}(r, r')\, G_{\omega\,\mathrm{refl}}^{(0)*}(r', r)
\end{aligned}
\qquad (7.72)
$$

everywhere apart from a thin surface layer of approximate thickness $k_F^{-1}$. Eq. (7.72) shows that $K_\omega(r, r')$ may again be expressed by a distribution function,

$$g_\omega(r, v; r') = g_{\omega\,\mathrm{dir}}(r, v; r') + g_{\omega\,\mathrm{refl}}(r, v; r'),\qquad (7.73)$$

where $g_{\omega\,\mathrm{dir}}(r, v; r')$ corresponds to the direct orbit and $g_{\omega\,\mathrm{refl}}(r, v; r')$ to the reflected orbit. The two distribution functions satisfy the Boltzmann equation (7.27) and the boundary conditions

$$
\begin{aligned}
g_{\omega\,\mathrm{dir}}(r, v_{\mathrm{out}}; r') &= 0 = g_{\omega\,\mathrm{refl}}(r, v_{\mathrm{in}}; r'),\\
g_{\omega\,\mathrm{refl}}(r, v_{\mathrm{out}}; r') &= g_{\omega\,\mathrm{dir}}(r, v_{\mathrm{in}}; r'),
\end{aligned}
\qquad (7.74)
$$

in a surface point $r$. The velocity vectors, $v_{\mathrm{out}}$ and $v_{\mathrm{in}}$, are, in the second line, related by Eq. (6.2). It follows from Eq. (7.74) that the righthand side of Eq. (7.73) indeed satisfies Eq. (6.1).

We emphasize that $K_\omega(r, r')$ may, in the immediate vicinity of the surface, not be expressed as an integral over a distribution function satisfying a Boltzmann equation. The kernel $K_\omega(r, r')$ is an extrapolation of the actual microscopic kernel (2.16) through a surface layer of approximate thickness $k_F^{-1}$. The method of the correlation function does not work at all if not all linear dimensions of a sample are big as compared to this length (which is of the order of the lattice constant). It is also clear that such very thin samples behave quite differently from a more physical point of view. It is perhaps noteworthy in this context that also other critical lengthes have to be big as compared to $k_F^{-1}$; cf. Eqs. (7.1), (7.31), and (7.54).

Considerations regarding boundary conditions in the microscopic theory of super-conductivity appear to have first been published by *Abrikosov* [31]. As pointed out by *Hu* and *Korenman* [32], his analysis is not correct in the presence of a magnetic field, since he multiplies the whole Green's function by a common phase factor instead of multi-plying each term on the righthand side of Eq. (7.71) by a phase factor corresponding to the respective orbit.

## 8. Proofs for Electrons in a Periodic Potential*

### a) General Formulations

The electrons in a metal are not free. The proofs of the Boltzmann equation and of the boundary condition for specular reflexion given in the previous section for free electrons, therefore, only concern a simple model but not a realistic metal. In this section we shall try to extend the proofs given in Section 7 to the model of electrons in a periodic potential [6]. Unfortunately, it will turn out that the proofs are less satisfactory than in the free electron case. Though the simple rule for spherical Fermi surfaces: "Do the calculations for free electrons and replace the electronic mass, $m$, by the effective mass, $m^*$" will be confirmed by our treatment, the logical basis for this rule does not seem fully conclusive.

In the absence of a magnetic field and of impurities, the matrix element of the one-electron Hamiltonian is now given by

$$\langle r|h|\,r''\rangle = \left(-\frac{1}{2m}\left(\frac{\partial}{\partial r}\right)^2 + V(r)\right)\delta(r-r'').\qquad(8.1)$$

The potential, $V(r)$, is assumed to possess the exact lattice periodicity. The solutions to the Schrödinger equation (3.7) then show a band structure. For each band, they may be chosen as Bloch functions and may be labeled by a wave number (or quasi-momentum) vector, $k$. The wave functions shall be denoted by $w_k(r)$. The quantity $\eta_m$ in Eq. (3.7) shall be called $\eta(k)$. We assume explicitly that the conduction electrons occupy a single band. All electronic wave functions and all values of $\eta(k)$ shall in the following belong to this conduction band.

---

* The authors feel that Section 8 is neither easy to read nor fully satisfactory. The readers not willing to dwell in the intricacies of Bloch functions, Wannier functions, various approximation methods, etc. are, therefore, advised to skip this section and to "believe" the Boltzmann equation.

Bloch functions, $w_k(r)$, may be expressed by Wannier functions, $a(r)$, in the following way,

$$w_k(r) = \sum_n \exp(i k \cdot n) a(r - n). \qquad (8.2)$$

The summation is extended over all lattice points, $n$. There are different Wannier functions for different bands. They are orthogonal to each other. In the following, by $a(r)$ we always mean the Wannier function of the conduction band. It is

$$\int a^*(r - n) a(r - n') \, d^3r = \Omega_0 \delta_{nn'}, \qquad (8.3)$$

if both Wannier functions belong to the same band (in particular to the conduction band). The normalization integral of the Wannier function has, for the sake of convenience, been put equal to the volume, $\Omega_0$, of the unit cell of the lattice. This convention leads to the usual normalization of the Bloch functions.

The Wannier function of the conduction band is not defined unambiguously since $w_k(r)$ in Eq. (8.2) may be multiplied by an arbitrary phase factor. In the following, it is sometimes assumed that the Wannier function is a real function. It is further assumed that $a(r)$ is localized around the origin. Then $a(r - n)$ is localized around the lattice point $n$. But it should not be forgotten that the Wannier function has a long oscillating tail.

It is seen from Eq. (3.9) that mainly one-electron states in the neighbourhood of the Fermi surface ($\eta = 0$) contribute to the Green's function, $G_\omega^{(0)}(r, r')$. Keeping only the wave functions of the conduction electrons on the righthand side and introducing Wannier functions with the help of Eq. (8.2), we obtain

$$G_\omega^{(0)}(r, r') \approx \sum_{nn'} a(r - n) \, \mathscr{G}_\omega^{(0)}(n, n') \, a^*(r' - n'), \qquad (8.4)$$

where $\mathscr{G}_\omega^{(0)}(n, n')$ is, in an infinite metal, given by

$$\mathscr{G}_\omega^{(0)}(n, n') = \frac{1}{(2\pi)^3} \int \frac{\exp[i k \cdot (n - n')]}{i\omega - \eta(k)} \, d^3k. \qquad (8.5)$$

The integration is extended, not over the whole $k$ space, but over a Brillouin zone. Since the concept of a conduction band is valid also in finite samples and in the presence of impurities or of a magnetic field, or of both, Eq. (8.4) is assumed to be applicable with the same Wannier function also under such more general conditions.

The kernel, $K_\omega(r, r')$, may be calculated from Eq. (8.4) with the help of Eq. (2.16). The Bloch functions introduce a kind of modulation with the periodicity of the lattice. The same periodic modulation will occur in $\Delta(r)$ as a consequence of the gap equation (2.15). Since the characteristic lengthes in the theory of superconductivity ($\xi_\omega$, $\zeta_\omega$, electronic mean free path) are much bigger than the lattice constant it seems reasonable to perform averages with respect to this small scale modulation. The way in which this will be done is, unfortunately, not fully satisfactory.

Eqs. (2.16), (8.4), and (8.3) suggest that, for real Wannier function, a reasonable spatial average of $K_\omega(r, r')$ is given by

$$\mathscr{K}_\omega(n, n') = \mathscr{G}_\omega^{(0)}(n, n')\, \mathscr{G}_\omega^{(0)*}(n', n)\,. \tag{8.6}$$

If the averaging procedure is also performed on the pair potential, $\Delta(r)$, the gap equation (2.15) might approximately (with an unknown error) be replaced by

$$\bar{\Delta}(n) = z g\, \Omega_0 \sum_{n'} T \sum_\omega{}' \mathscr{K}_\omega(n, n')\, \bar{\Delta}(n')\,, \tag{8.7}$$

where $z$ is a factor of order unity which results from the fact that an averaging of the individual factors on the righthand side of Eq. (2.15) is actually not legitimate. Whereas Eq. (8.7) perhaps meets with some doubts, it seems safe to replace the summation by an integration,

$$\bar{\Delta}(r) = z g \int T \sum_\omega{}' \mathscr{K}_\omega(r, r')\, \bar{\Delta}(r')\, \mathrm{d}^3 r'\,. \tag{8.8}$$

The righthand side has a meaning only if the definition of the averaged kernel can be extended also to space points not being lattice points.

Within Section 8 we shall stick to the notations just introduced. In all later applications, however, the averaging procedure shall not be exhibited. We shall again write $\Delta(r)$ instead of $\bar{\Delta}(r)$, and $K_\omega(r, r')$ instead of $\mathscr{K}_\omega(r, r')$. The factor $z$ shall be incorporated into a new, renormalized, coupling constant, $g$. Then Eq. (2.15) is again formally valid.

*b) Magnetic Field and Impurities Absent*

We want to prove that $\mathscr{K}_\omega(n, n')$ [or $\mathscr{K}_\omega(r, r')$] as defined by Eq. (8.6) may be expressed by a distribution function, $g_\omega(r, k; r')$, in the form

$$\mathscr{K}_\omega(r, r') = \int g_\omega(r, k; r')\, \delta[\eta(k)]\, \mathrm{d}^3 k = \oint g_\omega(r, k; r')\, \frac{\mathrm{d}S}{v(k)} \tag{8.9}$$

and that the distribution function satisfies a Boltzmann equation. Eq. (8.9) is identical with Eqs. (4.32) and (4.50) apart from some slight change in notation. $\mathrm{d}S$ is the surface element on the Fermi surface and $v(k)$ is the modulus of $v(k)$ [Eq. (4.40)]. The integration in the last integral is extended over the Fermi surface. For spherical Fermi surface (or, more exactly, spherical energy surfaces also in the neighbourhood of the Fermi surface), Eq. (8.9) reduces to

$$\mathcal{K}_\omega(r, r') = m^{*2} v \oint g_\omega(r, v; r') \, d\Omega \,, \tag{8.10}$$

where $m^*$ is the effective mass on the Fermi sphere as defined by Eq. (4.44). Eq. (8.10) is essentially identical with Eq. (4.55).

In the absence of a magnetic field and of impurity atoms, $\mathcal{G}_\omega^{(0)}(n, n')$ is, in a sufficiently extended metal, given by Eq. (8.5). It follows from Eq. (8.6) that it is

$$\mathcal{K}_\omega(r, r') = \frac{1}{(2\pi)^6} \int \frac{\exp[i(k + k') \cdot (r - r')]}{(\eta(k) - i\omega)(\eta(k') + i\omega)} \, d^3k \, d^3k'. \tag{8.11}$$

We shall first give a simple formal proof of Eq. (8.9), where $g_\omega(r, k; r')$ now satisfies the Boltzmann (Liouville) equation

$$\left(2|\omega| + v(k) \cdot \frac{\partial}{\partial r}\right) g_\omega(r, k; r') = \frac{1}{(2\pi)^2} \delta(r - r'). \tag{8.12}$$

Here, $v(k)$ is the velocity vector defined in Eq. (4.40) with the momentum vector $k$ being situated on the Fermi surface. Since the proof will be based upon some approximations whose validity is difficult to judge, we shall later on study the same problem in a different way.

Since the exponential in Eq. (8.11) is a rapidly oscillating function for not too small values of $|r - r'|$, we expect that only vectors $k + k'$ of sufficiently small modulus will contribute appreciably to the final result. Therefore, we put

$$k + k' = \kappa \tag{8.13}$$

and expand

$$\eta(k') = \eta(k) - v(k) \cdot \kappa + \cdots . \tag{8.14}$$

Here use has been made of Eqs. (4.40) and (5.28). The expansion shall be cut off after the two terms written down. Further we make use of Eq. (4.49). We assume $\omega > 0$, but the final result is also correct for $\omega < 0$. First the $\eta$ integral shall be done. It may, for sufficiently small values of $\omega$, be extended from $-\infty$ to $+\infty$ and leads to

$$\mathcal{K}_\omega(r, r') = \frac{1}{(2\pi)^5} \oint \frac{dS}{v(k)} \int \frac{\exp[i\kappa \cdot (r - r')]}{2\omega + iv(k) \cdot \kappa} \, d^3\kappa \tag{8.15}$$

with integration over the Fermi surface. Eq. (8.9) is in this way obtained with $g_\omega(r, k; r')$ given by

$$g_\omega(r, k; r') = \frac{1}{(2\pi)^5} \int \frac{\exp[i\kappa \cdot (r - r')]}{2\omega + iv(k) \cdot \kappa} \, d^3\kappa . \tag{8.16}$$

Eq. (8.12) follows immediately if the $\kappa$ integral is extended, not over a Brillouin zone, but over the whole $\kappa$ space and if it is remembered that $\omega$ was chosen as a positive quantity.

It is also possible to evaluate the righthand side of Eq. (8.5) directly for $|n - n'|$ (or $|r - r'|$) being big as compared to the lattice constant. The details of the calculation shall be given below in small print. The result is

$$\mathcal{G}_\omega^{(0)}(r, r')$$

$$= \left\{ \begin{matrix} -i \\ +i \end{matrix} \right\} \frac{1}{2\pi} {}_{+}\sqrt{\left| \frac{\det(m)}{\sum\limits_{ij} m_{ij}(x_i - x_i')(x_j - x_j')} \right|} \exp\left[ i k \cdot (r - r') - \frac{\omega}{v} |r - r'| \right]$$

$$(8.17)$$

for $\omega > 0$. The various signs and factors in the first curly bracket will be explained below. $k$ is that point on the Fermi surface for which $v(k)$ is parallel to $r - r'$. $v$ is again the modulus of $v(k)$. Also the mass tensor, which has been introduced in Section 4d, has to be evaluated at this point $k$. Eq. (8.6) leads from Eq. (8.17) to

$$\mathcal{K}_\omega(r, r') = \frac{1}{(2\pi)^2} \left| \frac{\det(m)}{\sum\limits_{ij} m_{ij}(x_i - x_i')(x_j - x_j')} \right| \exp\left[ -\frac{2|\omega|}{v} |r - r'| \right]. \quad (8.18)$$

Eq. (8.18) is fully identical with the classical result (4.56). It has already been shown in Section 4d that the righthand side of Eq. (8.18) may be obtained from the distribution function (4.52), which evidently satisfies Eq. (8.12).

It should be emphasized that Eqs. (8.17) and (8.18) are in general not correct for small values of $|r - r'|$. The short-distance behaviour of $\mathcal{K}_\omega(r, r')$ is, however, of no practical relevance since $\mathcal{K}_\omega(r, r')$ has, in Eq. (8.8), to be multiplied by a slowly varying function before an integration is performed with respect to the variable $r'$.

If there are several points $k$ on the Fermi surface for which $v(k)$ is parallel to $r - r'$, the righthand side of Eq. (8.17) has to be replaced by a sum of the corresponding contributions. The same is true for the righthand side of Eq. (8.18) if the oscillating interference terms are discarded which is again not permitted for too small values of $|r - r'|$. If the interference terms are left out, Eqs. (8.9) and (8.12) are still valid but $g_\omega(r, k; r')$ is now a non-vanishing function in all points $k$ for which $v(k)$ is parallel to $r - r'$. This problem does evidently not arise for a spherical Fermi surface.

We again assume $\omega > 0$. A similar treatment is possible for $\omega < 0$. The result is in accordance with Eq. (2.21). For sufficiently small values of $|\omega|$, only the neighbourhood of the Fermi surface contributes appreciably to the integral in Eq. (8.5). Because of the oscillating exponential, even only the neighbourhood of those points $k$ on the Fermi surface contributes for which the surface normal [or $v(k)$] is parallel or antiparallel to $r - r'$. As we shall see below, only the vicinity of $k$ with $v(k)$ parallel (and not antiparallel) to $r - r'$ does really contribute for $\omega > 0$.

We choose the orientation of a Cartesian coordinate system in such a way that $r - r'$ is parallel to the positive $z$ direction and that $(1/m)_{xy}$ vanishes in the point $k$. Then Eq. (8.5) may be written as

$$\mathscr{G}_{\omega}^{(0)}(r, r') = \frac{1}{(2\pi)^3} \int \frac{\exp[ik'_z |r - r'|]}{i\omega - \eta(k')} d^3 k' . \tag{8.19}$$

The function $\eta(k')$ is now expanded in the neighbourhood of the fixed point $k$ with the result

$$\eta(k') = v(k'_z - k) + \frac{1}{2}\left(\frac{1}{m}\right)_{xx} k'^2_x + \frac{1}{2}\left(\frac{1}{m}\right)_{yy} k'^2_y + \cdots . \tag{8.20}$$

Here, $k$ is the $z$ component of the vector $k$. $v$ is the $z$ component of the vector $v(k)$ whose $x$ and $y$ components vanish. Both $v$ and the reciprocal mass tensor shall be evaluated in the point $k$. Eq. (8.20) may be solved for $k'_z$,

$$k'_z = k + \frac{\eta}{v} - \frac{1}{2v}\left(\frac{1}{m}\right)_{xx} k'^2_x - \frac{1}{2v}\left(\frac{1}{m}\right)_{yy} k'^2_y + \cdots . \tag{8.21}$$

This expression is now inserted into the exponential in Eq. (8.19) which then factorizes. The individual factors depend only upon one of the quantities $\eta, k'_x, k'_y$. If we further put

$$d^3 k' = \frac{1}{v'_z} d\eta \, dk'_x \, dk'_y \tag{8.22}$$

and replace $v'_z$ by $v$ [the $z$ component of $v(k)$], the integrals may approximately be done.
    With $\omega > 0$ it is

$$\int_{-\infty}^{+\infty} \frac{\exp\left[i\frac{\eta}{v}|r - r'|\right]}{i\omega - \eta} d\eta = -2\pi i \exp\left[-\frac{\omega}{v}|r - r'|\right] \tag{8.23}$$

for $v > 0$ (classical motion from $r'$ to $r$). The righthand side vanishes for $v < 0$ [$v(k)$ and $r - r'$ antiparallel to each other]. The result

$$\int_{-\infty}^{+\infty} \exp[-i\alpha k'^2] \, dk' = \left\{\begin{matrix} e^{-\frac{i\pi}{4}} \\ e^{+\frac{i\pi}{4}} \end{matrix}\right\} \sqrt{\frac{\pi}{|\alpha|}} \quad \text{for} \quad \begin{cases} \alpha > 0, \\ \alpha < 0, \end{cases} \tag{8.24}$$

is also obtained by an integration in the complex plane. The path of integration is rotated by an angle of $-\frac{\pi}{4}$ or $+\frac{\pi}{4}$ depending upon the sign of $\alpha$. In this way, the integrand becomes real and leads to a well known integral. The collection of all factors essentially gives Eq. (8.17). The mass tensor element $m_{zz}$ is introduced with the help of Eq. (4.59). Finally, the orientation of the coordinate system is chosen in an arbitrary way.

The various signs and factors in the first curly bracket on the righthand side of Eq. (8.17) result from the two possible factors on the righthand side of Eq. (8.24). The upper sign $(-)$ is correct in Eq. (8.17) if both $(1/m)_{xx}$ and $(1/m)_{yy}$ are positive quantities. The lower sign $(+)$ holds if both elements of the reciprocal mass tensor are negative. The factor $-i$ finally is true if both elements have opposite signs. The whole analysis breaks down if $(1/m)_{xx}$ or $(1/m)_{yy}$ vanishes.

## c) Quasi-Classical Approximation

The Green's function $\mathscr{G}_\omega^{(0)}(r, r')$ may, in the absence of impurity atoms, also be calculated with the help of a quasi-classical approximation [6]. Though the derivation to be given below is not fully satisfactory it essentially leads back to Eq. (8.18), if there is no magnetic field. This agreement may be regarded as an argument in favour of the quasi-classical approximation. The method may then with some confidence be extended to the problem of electronic motion in a magnetic field. The same method of quasi-classical approximation will also be applied to the problem of impurity scattering further below (Section 7 d).

We start with a few general considerations (cf., e.g., *Ziman* [33]). $\eta(k)$, the energy in the conduction band measured with respect to the Fermi energy, may be expanded as follows,

$$\eta(k) = \sum_n \exp(-i\,k \cdot n)\,\eta_n . \tag{8.25}$$

The summation is extended over all lattice points, $n$, as in Eq. (8.2). The quantities $\eta_n$, which are real as a consequence of Eq. (4.24), are uniquely defined by this relation. They appear in the equation

$$\left(-\frac{1}{2m}\left(\frac{\partial}{\partial r}\right)^2 + V(r) - \mu\right) a(r) = \sum_n \eta_n a(r - n) \tag{8.26}$$

satisfied by the Wannier function, $a(r)$. $V(r)$ is the periodic potential introduced in Eq. (8.1).

After this preparation, we may attack the problem of calculating the Green's function. $G_\omega^{(0)}(r, r')$ is the unique solution to the differential equation

$$\left(i\omega + \mu + \frac{1}{2m}\left(\frac{\partial}{\partial r}\right)^2 - V(r)\right) G_\omega^{(0)}(r, r') = \delta(r - r'). \tag{8.27}$$

On the lefthand side, we use the ansatz (8.4). The delta function on the righthand side is replaced by

$$\delta(r - r') \rightarrow \frac{1}{\Omega_0} \sum_n a(r - n)\, a^*(r' - n), \tag{8.28}$$

which is that part of the completeness relation for Wannier functions which only contains the Wannier function of the conduction band. With the help of Eq. (8.28) we obtain the infinite set,

$$i\omega \mathscr{G}_\omega^{(0)}(n, n') - \sum_m \eta_{n-m} \mathscr{G}_\omega^{(0)}(m, n') = \frac{1}{\Omega_0} \delta_{nn'}, \tag{8.29}$$

of equations for $\mathscr{G}_\omega^{(0)}(n, n')$. Here, $\mathscr{G}_\omega^{(0)}(n, n')$ is defined on pairs of lattice points only. We try again to interpolate it by a function, $\mathscr{G}_\omega^{(0)}(r, r')$, of continuous arguments. It satisfies the equation

$$\left[i\omega - \eta\left(-i\frac{\partial}{\partial r}\right)\right]\mathscr{G}_\omega^{(0)}(r, r') = \delta(r - r'),\tag{8.30}$$

where $\eta\left(-\dfrac{\partial}{\partial r}\right)$ is obtained from Eq. (8.25) through the replacement

$$k \to -i\frac{\partial}{\partial r}\tag{8.31}$$

and where use has been made of the fact that $\exp(-n \cdot \partial/\partial r)$ is the operator of a translation by $-n$. The righthand side of Eq. (8.30) is the correct extension of the righthand side of Eq. (8.29) to continuous arguments. Eq. (8.30) shall now be solved by a suitable adaption of the quasi-classical approximation developed in Section 7a.

We assume that, for given $r$ and $r'$ and given value of $\eta(k)$, there is only one vector $k$ for which $v(k)$ [cf. Eq. (4.40)] is parallel to $r - r'$. The corresponding momentum shall, for $r'$ kept fixed, be denoted by $k(r)$. We then introduce the function

$$S(r, r') = k(r) \cdot (r - r'),\tag{8.32}$$

which satisfies

$$\frac{\partial S(r, r')}{\partial r} = k(r)\tag{8.33}$$

as a consequence of Eq. (4.40) or of the principle of least action. We finally make the ansatz

$$\mathscr{G}_\omega^{(0)}(r, r') = f_\omega(r)\exp[iS(r, r')]\tag{8.34}$$

for $\mathscr{G}_\omega^{(0)}(r, r')$, provided that $\omega$ is positive; cf. Eq. (7.12). It is

$$\eta\left(-i\frac{\partial}{\partial r}\right)\mathscr{G}_\omega^{(0)}(r, r') = \exp[iS(r, r')]\eta\left[k(r) - i\frac{\partial}{\partial r}\right]f_\omega(r)\tag{8.35}$$

as a consequence of the relation

$$-i\frac{\partial}{\partial r}\exp[iS(r, r')]f_\omega(r) = \exp[iS(r, r')]\left(k(r) - i\frac{\partial}{\partial r}\right)f_\omega(r),\tag{8.36}$$

which itself follows from Eq. (8.33). Therefore, $f_\omega(r)$ satisfies the equation

$$\left[i\omega - \eta\left[k(r) - i\frac{\partial}{\partial r}\right]\right]f_\omega(r) = \delta(r - r').\tag{8.37}$$

Now we regard the gradient as a small quantity and make a Taylor expansion,

$$\eta \left[ k(r) - i \frac{\partial}{\partial r} \right]$$

$$= \eta(k) - \frac{i}{2} \left( v(r) \cdot \frac{\partial}{\partial r} + \frac{\partial}{\partial r} \cdot v(r) \right) - \frac{1}{2} \sum_{ij} \left( \frac{1}{m} \right)_{ij} \frac{\partial}{\partial x_i} \frac{\partial}{\partial x_j} + \cdots . \tag{8.38}$$

$v(r)$ is the velocity vector defined in Eq. (4.40). It depends upon $r$ since $v$ shall be chosen parallel to $r - r'$. The reciprocal mass tensor has been defined in Eq. (4.41). Its components depend upon $k$ and so upon $r$. The term linear in the gradient has been symmetrized. It should, however, not be concealed that this simple symmetrization is not a consequence of Eq. (8.25). The term bilinear in the derivatives has not been symmetrized, since this term shall only be applied in a very approximate manner by neglecting the $r$ dependence of the reciprocal mass tensor altogether.

Eq. (8.38) is now inserted into Eq. (8.37). In the spirit of the quasi-classical approximation of Section 4a, we first put $\eta$ equal to zero so that the classical orbit is executed with Fermi energy. Further, we postulate

$$v(r) \cdot \frac{\partial f_\omega(r)}{\partial r} + \frac{1}{2} \frac{\partial v(r)}{\partial r} f_\omega(r) + \omega f_\omega(r) = 0 , \tag{8.39}$$

which is completely analogous to Eq. (7.15). It is solved in the same way with the result

$$f_\omega(r) = \frac{\text{const}}{|r - r'|} \exp \left[ - \frac{\omega |r - r'|}{v(r)} \right]. \tag{8.40}$$

The constant is determined by postulating

$$\frac{1}{2} \sum_{ij} \left( \frac{1}{m} \right)_{ij} \frac{\partial}{\partial x_i} \frac{\partial}{\partial x_j} f_\omega(r) = \delta(r - r') \tag{8.41}$$

and differentiating only the denominator in Eq. (8.40). The final result is

$$f_\omega(r) = \mp \frac{1}{2\pi} \sqrt{\frac{\det(m)}{\sum_{ij} m_{ij}(x_i - x_i')(x_j - x_j')}} \exp \left[ - \frac{\omega |r - r'|}{v} \right], \tag{8.42}$$

where the upper sign holds for positive definite and the lower sign for negative definite mass tensor. The method breaks down for an indefinite mass tensor. The result is identical with Eq. (8.17). But the original derivation of Eq. (8.17) was also valid for an indefinite mass tensor.

Eq. (8.41) is, with the $r$ dependence of the reciprocal mass tensor being neglected, solved by first transforming the mass tensor on principal axes and then choosing new units of length on each axis so that the differential operator, $\sum_{ij} \left(\dfrac{1}{m}\right)_{ij} \dfrac{\partial}{\partial x_i} \dfrac{\partial}{\partial x_j}$, becomes equal to the Laplacian (for positive definite mass tensor) or to its negative (for negative definite mass tensor). If the solution to this equation is then expressed by the original coordinates, a comparison of the result with Eq. (8.40) leads to Eq. (8.42).

Now we turn to the clean metal in a magnetic field. The behaviour of $G_\omega^{(0)}(r, r')$ under gauge transformations [Eq. (1.28)] is approximately correct also for $\mathscr{G}_\omega^{(0)}(r, r')$ as long as the gauge function, $U(r)$, is a slowly varying function on the atomic scale. We therefore conjecture that Eq. (8.30) has to be replaced by

$$\left[i\omega - \eta\left[-i\frac{\partial}{\partial r} + eA(r)\right]\right]\mathscr{G}_\omega^{(0)}(r, r') = \delta(r - r'). \qquad (8.43)$$

We refer to the literature [34, 35] for approximate derivations of this equation.

Eq. (8.43) is again solved in quasi-classical approximation. We introduce the canonical momentum, $p$, which is related to the kinetic momentum, $k$ by

$$p = k - eA(r); \qquad (8.44)$$

cf. Eq. (7.9). We regard

$$\eta(k) = \eta(p + eA(r)) =: \eta(p, r) \qquad (8.45)$$

as the Hamiltonian of the classical motion of a particle. The classical orbit will in general be curved. We again introduce the action integral

$$S(r, r') = \int_{r'}^{r} p(s) \cdot ds \qquad (8.46)$$

with integration along the classical orbit of energy $\eta$. With the ansatz (8.34) we obtain the same differential equation for $f_\omega(r)$ as without a magnetic field. The only difference is that the classical orbits are no longer straight lines. If, however, the curvature may be neglected on a length of the order of $v/|\omega|$, we again obtain Eq. (8.42) apart from a phase factor containing the vector potential. The kernel, $\mathscr{K}_\omega(r, r')$, is given by

$$\mathscr{K}_\omega(r, r') \qquad (8.47)$$

$$= \frac{1}{(2\pi)^2} \left| \frac{\det(m)}{\sum_{ij} m_{ij}(x_i - x_i')(x_j - x_j')} \right| \exp\left[-2ie\int_{r'}^{r} A(s) \cdot ds - \frac{2|\omega|}{v}|r - r'|\right].$$

A comparison with Eq. (8.18) shows that Eq. (8.9) holds with a distribution function, $g_\omega(r, k; r')$, satisfying the Boltzmann equation

$$(2|\omega| + v(k) \cdot \tilde{\partial}) \, g_\omega(r, k; r') = \frac{1}{(2\pi)^2} \, \delta(r - r') \,. \tag{8.48}$$

The symbol $\tilde{\partial}$ has been explained in Eq. (5.16).

The condition for the validity of the replacement of the curved orbit by a straight line shall not be discussed for Fermi surfaces of arbitrary shape. For a spherical Fermi surface, however, this condition is still given by Eq. (7.22) with the electronic mass, $m$, replaced by the effective mass, $m^*$.

### d) Impurity Scattering

The treatment of the scattering of conduction electrons by impurity atoms proceeds along the same lines as in Sections 7b and 7c. The potential produced by the impurities acts in real space. The basic formulations have, therefore, to be made in terms of the Green's functions $G_\omega^{(0)}(r, r')$ and not of the functions $\mathscr{G}_\omega^{(0)}(r, r')$. We again assume $\omega > 0$.

Eq. (7.37) remains essentially unaltered,

$$\langle G_\omega^{(0)}(r, r') \rangle$$
$$= G_\omega^{(0\,0)}(r, r') + \int G_\omega^{(0\,0)}(r, s) \, n(s) \, w(s, s') \, \langle G_\omega^{(0)}(s, s') \rangle \, \langle G_\omega^{(0)}(s', r') \rangle \, \mathrm{d}^3 s \, \mathrm{d}^3 s' \,. \tag{8.49}$$

We only discard the $\bar{v}$ term from the very beginning and write $w(s, s')$ instead of $w(s - s')$. The average,

$$w(s, s') = \int v(s - t) \, f(t) \, v(s' - t) \, \mathrm{d}^3 t \,, \tag{8.50}$$

might now contain an additional non-negative factor, $f(t)$, which has the periodicity of the lattice: an impurity atom preferentially sits on certain positions in the unit cell of the lattice. $f(t)$ is normalized in such a way that the integral of this function over a unit cell of the lattice equals its volume, $\Omega_0$. The impurity concentration, $n(s)$, is regarded as a function which varies at most on a macroscopic scale.

Eq. (7.40) is, for neighbouring points $s'$ and $s$, replaced by

$$\langle G_\omega^{(0)}(s, s') \rangle \approx - \frac{i\pi}{(2\pi)^3} \int w_{k'}(s) \, w_{k'}^*(s') \, \delta[\eta(k')] \, \mathrm{d}^3 k' \,. \tag{8.51}$$

This equation, which has to be checked on the final result, follows, in the absence of impurities, from Eqs. (8.4), (8.5), and (8.2). The principal-value integral, which had been retained in Eq. (7.40), is here discarded from the very beginning. Instead of Eq. (7.41), we claim

$$\langle G_\omega^{(0)}(s', r')\rangle = \frac{w_k(s')}{w_k(s)} \langle G_\omega^{(0)}(s, r')\rangle . \tag{8.52}$$

This equation, in the absence of impurity atoms, follows from Eqs. (8.4), (8.17), and (8.2). $k$ is that momentum vector on the Fermi surface for which $v(k)$ is parallel to $s - r'$. We shall not regard the situation when there are several values of $k$ satisfying this condition.

Making now use of the differential equation (8.27) for $G_\omega^{(0\,0)}(r, r')$ [which in that equation had been called $G_\omega^{(0)}(r, r')$], we obtain

$$\left(i\omega + \frac{i}{2} n(r) v(k) \sigma(k, r) + \frac{1}{2m} \left(\frac{\partial}{\partial r}\right)^2 - V(r)\right) \langle G_\omega^{(0)}(r, r')\rangle = \delta(r - r') \tag{8.53}$$

with $v(k) \sigma(k, r)$ given by

$$v(k) \sigma(k, r) \tag{8.54}$$

$$= \frac{1}{|w_k(r)|^2} \frac{1}{(2\pi)^2} \int w_k^*(r) w_{k'}(r) w(r, s') w_{k'}^*(s') w_k(s') \delta[\eta(k')] \, d^3 k' \, d^3 s' .$$

The righthand side of the last equation shows the periodicity of the lattice if $k$ is kept fixed. Actually, the vector $k$ is a slowly varying function of $r$ for sufficiently big values of $|r - r'|$. The surprising first factor on the righthand side of Eq. (8.54), with the wave function in the denominator, will again disappear in the course of the calculations.

We go over from Eq. (8.53) to an equation for $\langle \mathscr{G}_\omega^{(0)}(n, n')\rangle$ which is, apart from the angular brackets, defined by Eq. (8.4). The same procedure that led from Eq. (8.27) to Eq. (8.29) now approximately gives

$$i\omega \langle \mathscr{G}_\omega^{(0)}(n, n')\rangle$$

$$- \sum_m \left[\eta_{n-m} - \frac{i}{2} n(n) v(k) \frac{1}{\Omega_0} \int a^*(r - n) \sigma(k, r) a(r - m) d^3 r\right] \langle \mathscr{G}_\omega^{(0)}(m, n')\rangle$$

$$= \frac{1}{\Omega_0} \delta_{nn'} . \tag{8.55}$$

The approximation made is that the actual $r$ dependence of $k$ has been neglected. This seems a reasonable approximation for sufficiently big values of $|n - n'|$ but it is quite wrong for small values of $|n - n'|$. Making further use of

$$\langle \mathscr{G}_\omega^{(0)}(m, n')\rangle = \exp[i k \cdot (m - n)] \langle \mathscr{G}_\omega^{(0)}(n, n')\rangle , \tag{8.56}$$

which corresponds to Eq. (7.41), and making the same transition as from Eq. (8.29) to Eq. (8.30), we finally arrive at

$$\left[i\omega + \frac{i}{2} n(r) v(k) \sigma(k) - \eta\left(-i\frac{\partial}{\partial r}\right)\right] \langle \mathscr{G}_\omega^{(0)}(r, r')\rangle = \delta(r - r') \quad (8.57)$$

with

$$v(k) \sigma(k) = \quad\quad\quad\quad\quad\quad\quad\quad\quad\quad\quad\quad\quad\quad\quad\quad (8.58)$$

$$= \frac{1}{\Omega_0} \int_{\Omega_0} d^3 t f(t) \int d^3 k' \, \delta[\eta(k')] \frac{1}{(2\pi)^2} |\int d^3 s w_{k'}^*(s) v(s-t) w_k(s)|^2.$$

The details will be given below in small print. In Eq. (8.58), the $t$ integration is restricted to a unit cell of the lattice, with volume $\Omega_0$. In Eq. (8.58), not only the strange denominator $|w_k(r)|^2$ has disappeared. It also is that relation for $\sigma(k)$ which would be obtained in Born approximation. It even includes an average over the lattice cell with the weight factor $f(t)$. The quasi-classical approximation may again be applied to Eq. (8.57). A comparison with the analogous problem in the absence of impurities shows, however, directly that

$$\langle \mathscr{G}_\omega^{(0)}(r, r')\rangle \langle \mathscr{G}_\omega^{(0)*}(r', r)\rangle = \oint g_\omega^{(\sigma)}(r, k; r') \frac{dS}{v(k)} \quad\quad (8.59)$$

is valid with the distribution function, $g_\omega^{(\sigma)}(r, k; r')$, satisfying the Boltzmann equation

$$\left(2|\omega| + v \cdot \frac{\partial}{\partial r} + n(r) v(k) \sigma(k)\right) g_\omega^{(\sigma)}(r, k; r') = \frac{1}{(2\pi)^2} \delta(r - r') . \quad (8.60)$$

Not only Eq. (8.2) has again been applied in the derivation of Eqs. (8.57) and (8.58). In the first step, Eq. (8.57) is obtained with

$$\sigma(k) = \sum_m \frac{1}{\Omega_0} \int a^*(r-n) \sigma(k, r) a(r-m) \exp[ik \cdot (m-n)] \, d^3 r. \quad (8.61)$$

If $\sigma(k, r)$ is regarded as a strictly periodical function of $r$ (with the periodicity of the lattice) the $r$ integration over the whole space may be replaced by an integration over the unit cell and a summation over all lattice points, $n$. Making again use of Eq. (8.2), we obtain

$$\sigma(k) = \frac{1}{\Omega_0} \int_{\Omega_0} |w_k(r)|^2 \sigma(k, r) \, d^3 r. \quad (8.62)$$

In the final step, the restriction on the $r$ integration is transferred to a restriction on the $t$ integration. Eq. (8.58) is obtained from Eq. (8.62) with the help of Eq. (8.54).

Now we have to study the kernel $K_\omega(r, r', r'')$ [Eq. (7.59)] for which Eq. (7.62) still holds with the only alteration that $w(s - s'')$ is again replaced by $w(s, s'')$ [Eq. (8.50)]. From Eq. (7.62) we then go over to an equation for the kernel

$$\mathcal{K}_\omega(n, n', n'') = \langle \mathcal{G}_\omega^{(0)}(n, n') \mathcal{G}_\omega^{(0)*}(n', n'') \rangle . \tag{8.63}$$

This equation is given by

$$\mathcal{K}_\omega(n, n', n'') = \langle \mathcal{G}_\omega^{(0)}(n, n') \rangle \langle \mathcal{G}_\omega^{(0)*}(n', n'') \rangle$$
$$+ \sum_{\substack{p\,p'' \\ m\,m''}} \langle \mathcal{G}_\omega^{(0)}(n, p) \rangle \langle \mathcal{G}_\omega^{(0)*}(p, n'') \rangle \tag{8.64}$$
$$\cdot \mathcal{K}_\omega(p, n', p'') \int a^*(s - m)\, a(s - p)\, a(s'' - p'')\, a^*(s'' - m'')$$
$$\cdot \exp[i k \cdot (m'' - m)]\, n(s)\, w(s, s'')\, \mathrm{d}^3 s\, \mathrm{d}^3 s''.$$

Eq. (8.4) and essentially Eq. (8.56) have been used in the derivation. $k$ is again that momentum on the Fermi surface for which the velocity vector is parallel to $n - n'$. The $m$ and $m''$ summations may be performed in Eq. (8.64) if use is made of Eq. (8.2).

In order to introduce the kernel $\mathcal{K}_\omega(n, n', n'')$, it is actually necessary first to introduce

$$K_\omega(r, r', r''', r'') = \langle G_\omega^{(0)}(r, r')\, G_\omega^{(0)*}(r''', r'') \rangle \tag{8.65}$$

for neighbouring points $r'$ and $r'''$ and then to go over to

$$\mathcal{K}_\omega(n, n', n''', n'') = \langle \mathcal{G}_\omega^{(0)}(n, n')\, \mathcal{G}_\omega^{(0)*}(n''', n'') \rangle \tag{8.66}$$

with the help of Eq. (8.4). Whereas $K_\omega(r, r', r''', r'')$ satisfies an integral equation analogous to Eq. (7.62), $K_\omega(n, n', n''', n'')$ obeys an equation analogous to Eq. (8.64). Eq. (8.64) itself is obtained by putting $n''' = n'$.

A slight generalization of Eq. (8.59) is given by

$$\langle \mathcal{G}_\omega^{(0)}(n, n') \rangle \langle \mathcal{G}_\omega^{(0)*}(n', n'') \rangle$$
$$= \oint g_\omega^{(\sigma)}(n, k; n') \exp[i k \cdot (n - n'')] \frac{\mathrm{d} S}{v(k)} . \tag{8.67}$$

This equation is quite analogous to Eq. (7.65). Inserting Eq. (8.67) into Eq. (8.64) we see that $\mathcal{K}_\omega(n, n', n'')$ may be written in the form

$$\mathcal{K}_\omega(n, n', n'') = \oint g_\omega(n, k; n') \exp[i k \cdot (n - n'')] \frac{\mathrm{d} S}{v(k)}, \tag{8.68}$$

where the so far unknown function $g_\omega(n, k; n')$ is the distribution function we are looking for. This follows from the fact that it is

$$\mathcal{K}_\omega(n, n', n) = \mathcal{K}_\omega(n, n'). \tag{8.69}$$

The function $g_\omega(n, k; n')$ satisfies the equation

$$g_\omega(n, k; n') = g_\omega^{(\sigma)}(n, k; n') + \sum_{p\,p''} g_\omega^{(\sigma)}(n, k; p)\, \mathcal{K}_\omega(p, n', p'')$$
$$\int w_k^*(s)\, a(s - p)\, a(s'' - p'')\, w_{-k}^*(s'')\, n(s)\, w(s, s'')\, \mathrm{d}^3 s\, \mathrm{d}^3 s'', \tag{8.70}$$

which follows from Eq. (8.64). As in the free electron case, $\mathcal{K}_\omega(p, n', p'')$ is, on the righthand side, now expressed by Eq. (8.68) where the variable of integration is called $k'$. It is again possible to apply Eq. (8.2).

In order to obtain the expected result it seems unavoidable to make use of the invariance under time reversal in the form

$$w^*_{-k}(s'') = w_k(s'') .\tag{8.71}$$

This equation is automatically satisfied if the Wannier function $a(r)$ is a real function. A number of manipulations which shall be explained below in small print then leads to

$$g_\omega(r, k; r') = g_\omega^{(\sigma)}(r, k; r')$$
$$+ (2\pi)^2 \int d^3s \, g_\omega^{(\sigma)}(r, k; s) \, n(s) \oint \frac{dS'}{v(k')} P(k, k') \, g_\omega(s, k'; r') ,\tag{8.72}$$

where $r$ and $r'$ are now regarded as continuous variables. Here, the transition probability, $P(k, k')$, is defined by

$$P(k, k') = \frac{1}{\Omega_0} \int_{\Omega_0} d^3t \, f(t) \frac{1}{(2\pi)^2} \left| \int d^3s \, w_k^*(s) \, v(s-t) \, w_{k'}(s) \right|^2 .\tag{8.73}$$

This is again the result expected in Born approximation. Eqs. (8.73) and (8.58) together satisfy Eq. (5.5). If now use is made of Eq. (8.60), we obtain the Boltzmann equation

$$\left( 2|\omega| + v \cdot \frac{\partial}{\partial r} + n(r) \, v(k) \, \sigma(k) \right) g_\omega(r, k; r')$$
$$- n(r) \oint P(k, k') \, g_\omega(r, k'; r') \frac{dS'}{v(k')} = \frac{1}{(2\pi)^2} \delta(r - r') .\tag{8.74}$$

It is identical with Eq. (5.7). We expect that an analogous treatment in the presence of a magnetic field will lead to the same Boltzmann equation apart from the replacement (5.16).

If Eq. (8.68) is, with the appropriate notational changes, inserted into the righthand side of Eq. (8.70), the summation with respect to $p''$ may be performed with the help of Eq. (8.2). Also here, Eq. (8.71) has to be applied. Then the $s$ integration is restricted to a unit cell of the lattice. If this restriction is compensated by a summation with respect to the lattice points, also this sum leads to a one-electron wave function. The restriction on the $s$ integration is then again transferred to a corresponding restriction on the $t$ integration. Finally, the discrete arguments, $n$, $p$, and $n'$ are replaced by continuous arguments, $r$, $s$, and $r'$, respectively, and the $p$ summation is performed as an $s$ integration divided by the volume of the unit cell, $\Omega_0$, (This last step is permitted if both $|n - p|$ and $|p - n'|$ are sufficiently big. It is certainly not correct if at least one of these lengthes is small.) In this way Eqs. (8.72) and (8.73) are obtained.

## 9. Paramagnetic Impurities

### a) Warning

If the electron-impurity interaction is spin dependent it is not possible to separate off the spin degree of freedom from the very beginning. The matrix element of the one-electron Hamiltonian is now given by

$$\langle r\alpha|h|\,\alpha''r''\rangle = \left[\left(-\frac{1}{2m}\left(\frac{\partial}{\partial r}\right)^2 - \mu\right)\delta_{\alpha\alpha''} + W_{\alpha\alpha''}(r)\right]\delta(r - r'') \quad (9.1)$$

with

$$W_{\alpha\alpha''}(r) = \sum_j \left(v_1(r - r_j)\,\delta_{\alpha\alpha''} + v_2(r - r_j)\,\sigma_{\alpha\alpha''} \cdot S_j\right), \quad (9.2)$$

provided that the electrons are regarded as free. $S_j$ is the impurity spin operator. A state of paramagnetic ordering of the impurity spins is assumed which shall be formulated in detail below. $\sigma_{\alpha\alpha''}$ is the Pauli spin matrix of the electron. The interaction potentials, $v_k(r)$, are assumed to be rotationally invariant.

The influence of paramagnetic impurities on superconductivity problems has been studied theoretically within the Gorkov scheme by *Abrikosov* and *Gorkov* [36]. Their method is essentially the same as in the case of non-magnetic impurities [30]. So, the interaction of an electron with an individual impurity atom is only treated up to second order of perturbation theory. It is not difficult also to generalize the derivation of the Boltzmann equation as given in Sections 7 and 8 to the case of paramagnetic impurities.

Before this will be done in a rather straightforward manner, a few remarks of more general nature seem in order. It has been known since the work of *Kondo* [37] that the interaction between the spins of a conduction electron and of an impurity atom is not treated properly in a low order perturbation calculation. Drastic effects occur if the interaction is taken into account in a correct way. This at least is true in the normal state. It was, however, to be expected that also the Abrikosov-Gorkov analysis of the superconducting state might be drastically incorrect because of the low order treatment of the electron-impurity interaction.

Work on the influence of magnetic impurities on superconductivity problems is still in progress at various places. From the work published so far (cf. *Zittartz* and *Müller-Hartmann* [38] for a list and critical survey of the literature), it seems that the results of *Abrikosov* and *Gorkov* are fairly reliable for attractive spin-spin interaction [$v_2(r)$ essentially negative] but might, for an appropriate choice of parameters, be greatly in error for repulsive spin-spin interaction*. We hope that also the

---

* We want to thank E. *Müller-Hartmann* for discussions on this subject.

derivation of the Boltzmann equation to be given in this section is tenable for attractive spin-spin interaction. If this were so, also the results to be derived later in Chapter II should be reliable. For repulsive interaction, however, some reserve towards the Boltzmann equation and the results derived from it seems in order. In any case, the reader should be aware of the problematics of the interaction between electron spins and impurity spins.

### b) Derivation of the Boltzmann Equation

We want to derive the modification of the Boltzmann equation due to paramagnetic impurities [4]. In order to avoid additional complications, which are not relevant for the present problem, the electrons shall be regarded as free. The Hamiltonian is, therefore, given by Eq. (9.1). It is expected that the result will be essentially the same also for electrons in a periodic potential.

It is again necessary to first calculate the impurity averaged Green's function, $\langle G^{(0)}_{\omega\alpha\alpha'}(r, r')\rangle$. Before any averaging is done, the Green's function $G^{(0)}_{\omega\alpha\alpha'}(r, r')$ is the solution to the equation

$$G^{(0)}_{\omega\alpha\alpha'}(r, r') = G^{(0\,0)}_{\omega}(r, r')\,\delta_{\alpha\alpha'} + \int G^{(0\,0)}_{\omega}(r, s)\,W_{\alpha\beta}(s)\,G^{(0)}_{\omega\beta\alpha'}(s, r')\,d^3 s \qquad (9.3)$$

which corresponds to Eq. (7.34). $G^{(0\,0)}_{\omega}(r, r')$ is still given by Eq. (4.10). A summation with respect to spin subscripts appearing twice is tacitly assumed in Eq. (9.3) and in later equations. $G^{(0)}_{\omega\alpha\alpha'}(r, r')$ does not only, as in the case of non-magnetic impurities, depend upon the particular choice of the positions $r_j$ of the impurity atoms. It also is, because of the impurity spin operators, an operator in the spin space of the impurity atoms.

The final averaging procedure has to be done not only with respect to the impurity positions but also with respect to the spin states. The spins of different impurities are uncorrelated in a state of paramagnetic ordering so that all averages containing spin operators of different impurities factorize. For paramagnetic ordering it further is

$$\langle S\rangle = 0 \qquad (9.4)$$

and

$$\langle S_i S_j\rangle = \delta_{ij}\,\frac{S(S+1)}{3}, \qquad (9.5)$$

if the impurity spin is equal to $S$. The subscripts in Eq. (9.5) refer to Cartesian components of the same spin. It follows from Eq. (9.5) that it is

$$\langle(\sigma_{\alpha\beta}\cdot S)\,(\sigma_{\beta'\gamma}\cdot S)\rangle = \sigma_{\alpha\beta}\cdot\sigma_{\beta'\gamma}\,\frac{S(S+1)}{3}. \qquad (9.6)$$

Averages containing more than two spin operators of the same impurity atom do not occur in the calculations.

The iterative solution to Eq. (9.3) is obtained in the same way as formerly was the iterative solution to Eq. (7.34). Also the averaging procedure is done in essentially the same manner. If the interaction of the electron with an individual impurity atom is treated up to second order, if contributions with crossing impurity lines are neglected, and if contributions leading to a one-electron potential are discarded from the very beginning, we end up with

$$\langle G^{(0)}_{\omega \alpha \alpha'}(r, r')\rangle = G^{(0\,0)}_{\omega}(r, r')\,\delta_{\alpha\alpha'}$$

$$+ \int G^{(0\,0)}_{\omega}(r, s)\,n(s)\left[w_1(s-s')\,\delta_{\alpha\beta}\delta_{\beta'\gamma} + w_2(s-s')\,\frac{S(S+1)}{3}\,\boldsymbol{\sigma}_{\alpha\beta}\cdot\boldsymbol{\sigma}_{\beta'\gamma}\right] \qquad (9.7)$$

$$\cdot\,\langle G^{(0)}_{\omega\beta\beta'}(s, s')\rangle\,\langle G^{(0)}_{\omega\gamma\alpha'}(s', r')\rangle\,\mathrm{d}^3s\,\mathrm{d}^3s'\,,$$

where use has been made of Eqs. (9.4) and (9.6). The analogous equation for non-magnetic impurities was given by Eq. (7.37). It is

$$w_1(s-s') := \int v_1(s-t)\,v_1(s'-t)\,\mathrm{d}^3t = \frac{1}{(2\pi)^3}\int |v_1(k)|^2\exp[i\boldsymbol{k}\cdot(s-s')]\,\mathrm{d}^3k\,,$$

$$w_2(s-s') := \int v_2(s-t)\,v_2(s'-t)\,\mathrm{d}^3t = \frac{1}{(2\pi)^3}\int |v_2(k)|^2\exp[i\boldsymbol{k}\cdot(s-s')]\,\mathrm{d}^3k\,,$$

$$(9.8)$$

in analogy with Eqs. (7.38) and (7.43). Here, the Fourier transforms, $v_1(k)$ and $v_2(k)$, are defined in accordance with Eq. (7.42).

An iterative solution to Eq. (9.7) leads to an averaged Green's function which is diagonal in spin space,

$$\langle G^{(0)}_{\omega\alpha\alpha'}(r, r')\rangle = \langle G^{(0)}_{\omega}(r, r')\rangle\,\delta_{\alpha\alpha'}\,. \qquad (9.9)$$

This result follows from

$$\boldsymbol{\sigma}_{\alpha\beta}\cdot\boldsymbol{\sigma}_{\beta\gamma} = 3\,\delta_{\alpha\gamma}\,. \qquad (9.10)$$

If the ansatz (9.9) is introduced into Eq. (9.7), we obtain

$$\langle G^{(0)}_{\omega}(r, r')\rangle$$

$$= G^{(0\,0)}_{\omega}(r, r') + \int G^{(0\,0)}_{\omega}(r, s)\,n(s)\,[w_1(s-s') + S(S+1)\,w_2(s-s')] \qquad (9.11)$$

$$\cdot\,\langle G^{(0)}_{\omega}(s, s')\rangle\,\langle G^{(0)}_{\omega}(s', r')\rangle\,\mathrm{d}^3s\,\mathrm{d}^3s'\,.$$

Here, use has again been made of Eq. (9.10).

Eq. (9.11) is fully equivalent to the corresponding equation for electrons in the presence of non-magnetic impurities, i.e., to Eq. (7.37). Therefore, also the same conclusions may be drawn. Eq. (7.57) still holds.

But $g_\omega^{(\sigma)}(r, v; r')$ now obeys the differential equation

$$\left(2|\omega| + v \cdot \frac{\partial}{\partial r} + nv\sigma_1\right) g_\omega^{(\sigma)}(r, v; r') = \delta(r - r'), \qquad (9.12)$$

where

$$\sigma_1 = \left(\frac{m}{2\pi}\right)^2 \oint \left(|v_1(k - k')|^2 + S(S + 1)|v_2(k - k')|^2\right) d\Omega \qquad (9.13)$$

is the total scattering cross section (including spin flip) in Born approximation. Both $k$ and $k'$ are momentum vectors on the Fermi sphere. The integration is extended over all directions of $k$.

The gap equation, in the spin dependent case, reads

$$\Delta(r)\,\delta_{\alpha\alpha'} = g \int d^3 r'\, T \sum_\omega{}' G_{\omega\alpha\beta}^{(0)}(r, r')\, \Delta(r')\, \tilde{G}_{\omega\beta\alpha'}^{(0)}(r', r). \qquad (9.14)$$

Here, $\tilde{G}_{\omega\alpha\alpha'}^{(0)}(r, r')$ is defined by

$$\tilde{G}_{\omega\alpha\alpha'}^{(0)}(r, r') = -\varepsilon_{\alpha\beta} G_{\omega\beta\beta'}^{(0)*}(r, r')\, \varepsilon_{\beta'\alpha'} \qquad (9.15)$$

with the matrix $\varepsilon_{\alpha\beta}$ given in Eq. (1.3). It follows from Eq. (9.3) that $\tilde{G}_{\omega\alpha\alpha'}^{(0)}(r, r')$ satisfies the integral equation

$$\tilde{G}_{\omega\alpha\alpha'}^{(0)}(r, r') = G_\omega^{(00)*}(r, r')\,\delta_{\alpha\alpha'} + \int G_\omega^{(00)*}(r, s)\,\tilde{W}_{\alpha\beta}(s)\,\tilde{G}_{\omega\beta\alpha'}^{(0)}(s, r')\, d^3 s \qquad (9.16)$$

with

$$\tilde{W}_{\alpha\beta}(s) = \sum_j \left(v_1(r - r_j)\,\delta_{\alpha\beta} - v_2(r - r_j)\,\sigma_{\alpha\beta} \cdot S_j\right) \qquad (9.17)$$

as a consequence of Eq. (1.4) and of

$$-\varepsilon_{\alpha\beta}\,\sigma_{\beta\beta'}^*\,\varepsilon_{\beta'\alpha'} = -\sigma_{\alpha\alpha'}. \qquad (9.18)$$

That the righthand side of Eq. (9.14) is really diagonal in spin space may be derived from symmetry properties of spin dependent Green's functions. For the present purpose it suffices to show that this is true after the impurity average has been performed. It will indeed turn out below that it is

$$K_{\omega\alpha\alpha'}(r, r') = \langle G_{\omega\alpha\beta}^{(0)}(r, r')\, \tilde{G}_{\omega\beta\alpha'}^{(0)}(r', r)\rangle = K_\omega(r, r')\,\delta_{\alpha\alpha'}. \qquad (9.19)$$

The gap equation is, therefore, still given by Eq. (2.15).

We again introduce the quantity

$$K_{\omega\alpha\alpha'}(r, r', r'') = \langle G_{\omega\alpha\beta}^{(0)}(r, r')\, \tilde{G}_{\omega\beta\alpha'}^{(0)}(r', r'')\rangle, \qquad (9.20)$$

which satisfies the integral equation

$$K_{\omega\alpha\alpha''}(r, r', r'') = \langle G_\omega^{(0)}(r, r')\rangle \, \langle G_\omega^{(0)*}(r', r'')\rangle \, \delta_{\alpha\alpha''} + \int \langle G_\omega^{(0)}(r, s)\rangle \, n(s)$$

$$\cdot \left[ w_1(s-s'') \, \delta_{\alpha\beta}\delta_{\beta''\alpha''} - w_2(s-s'') \, \frac{S(S+1)}{3} \, \boldsymbol{\sigma}_{\alpha\beta} \cdot \boldsymbol{\sigma}_{\beta''\alpha''} \right] \qquad (9.21)$$

$$\cdot K_{\omega\beta\beta''}(s, r', s'') \, \langle G_\omega^{(0)*}(s'', r'')\rangle \, d^3s \, d^3s''$$

in close analogy to Eq. (7.62). This equation is solved by a $K_{\omega\alpha\alpha''}(r, r', r'')$ which is diagonal in spin space,

$$K_{\omega\alpha\alpha''}(r, r', r'') = K_\omega(r, r', r'') \, \delta_{\alpha\alpha''} . \qquad (9.22)$$

We obtain

$$K_\omega(r, r', r'') = \langle G_\omega^{(0)}(r, r')\rangle \, \langle G_\omega^{(0)*}(r', r'')\rangle + \int \langle G_\omega^{(0)}(r, s)\rangle \, n(s) \qquad (9.23)$$

$$[w_1(s-s'') - S(S+1) \, w_2(s-s'')] \, K_\omega(s, r', s'') \, \langle G_\omega^{(0)*}(s'', r'')\rangle \, d^3s \, d^3s'' .$$

This equation is now fully equivalent to the corresponding Eq. (7.62) in the case of non-magnetic impurities. The same manipulations as in Section 7c and application of Eq. (9.12) lead to

$$\left( 2|\omega| + \boldsymbol{v} \cdot \frac{\partial}{\partial r} + nv\sigma_1 \right) g_\omega(r, \boldsymbol{v}; r')$$

$$- nv \oint \frac{d\sigma_2(\theta)}{d\Omega} \, g_\omega(r, \boldsymbol{v}'; r') \, d\Omega' = \frac{1}{(2\pi)^2} \, \delta(r - r') \qquad (9.24)$$

with the differential scattering cross section $d\sigma_2(\theta)/d\Omega$ given by

$$\frac{d\sigma_2(\theta)}{d\Omega} = \left( \frac{m}{2\pi} \right)^2 (|v_1(k-k')|^2 - S(S+1) \, |v_2(k-k')|^2). \qquad (9.25)$$

$\theta$ is the angle between the directions of the vectors $k$ and $k'$ which both lie on the Fermi sphere. In the presence of a magnetic field, the substitution (5.16) has to be made in Eq. (9.24).

If we now introduce the quantity

$$\sigma_2 = \oint \frac{d\sigma_2(\theta)}{d\Omega} \, d\Omega , \qquad (9.26)$$

it is not $\sigma_1 = \sigma_2$ but instead it is

$$\sigma_{sp} = \frac{\sigma_1 - \sigma_2}{2} = \left( \frac{m}{2\pi} \right)^2 S(S+1) \oint |v_2(k-k')|^2 \, d\Omega \geq 0 . \qquad (9.27)$$

$\sigma_{sp}$ is the spin flip cross section. The result that $\sigma_1$ and $\sigma_2$ are different quantities is a direct consequence of the fact that the Hamiltonian (9.1) is not time reversal invariant if the operation of time reversal is not also applied to the impurity spins. The intuitive picture that $g_\omega(r, v; r')$ is a distribution function of classical particles does, therefore, not work here even in the absence of a magnetic field. Indeed the number of particles is not conserved. It decreases monotonically as a consequence of $\sigma_{sp} > 0$. If the time dependent distribution function, $f(r, v; r'; t)$ is temporarily again introduced [cf. Eqs. (4.31) and (5.7)], we find

$$\frac{\partial}{\partial t} \int d^3 r \oint d\Omega \, f(r, v; r'; t) = - \int d^3 r \, nv\sigma_{sp} \oint d\Omega \, f(r, v; r'; t). \quad (9.28)$$

A consequence of the non-conservation of particles is that the sum rule (3.13) does not hold even for $A(r) \equiv 0$ but is replaced by

$$\int K_\omega(r, r') \, d^3 r = \frac{\pi N(r')}{(|\omega| + nv\sigma_{sp})}, \quad (9.29)$$

provided that $nv\sigma_{sp}$ is position independent in the sample. This statement follows immediately, if the considerations after Eq. (5.17) are repeated in the present case.

It will turn out in the end of Section 11c that the transition temperature lowering effect of paramagnetic impurities is a direct consequence of this non-conservation of particles.

The derivation of Eq. (9.7) is not correct. The impurity spins should be treated, not as operators, but as classical vectors in order to derive this equation.

# II. General Discussion. Approximation Methods

## 10. Survey of Chapter I

### a) Gap Equation

At a second kind transition between normal and superconducting state, Gorkov's self-consistency condition for the pair potential, $\Delta(r)$, (gap equation) reduces to a linear integral equation,

$$\Delta(r) = g \int d^3 r' \, T \sum_\omega{}' K_\omega(r, r') \, \Delta(r') . \tag{10.1}$$

The integral is extended over the metallic sample which is assumed to be electrically insulated against its surrounding. $g$ is the BCS coupling constant. It measures the strength of the phonon-mediated electron-electron interaction. $T$ is the absolute temperature. The $\omega$ summation is extended over the odd negative and positive multiples of $\pi T$. The dash on the summation sign indicates that this summation has to be cut off at $\omega = \pm \omega_D$ where $\omega_D$ is a frequency of the order of the Debye frequency. It is the task of the method of the correlation function to calculate the kernel, $K_\omega(r, r')$.

If a sample consists of several metals in electric contact, the coupling constant and the Debye frequency become position dependent. But they are still assumed to be constant inside each metal. The question arises as to how the $\omega$ cut-off should be done in this case. It seems suggestive to do this in a symmetrical manner through replacing Eq. (10.1) by

$$\Delta(r) = g(r) \int d^3 r' \, T \sum_\omega{}' \theta_\omega(r, r') \, K_\omega(r, r') \, \Delta(r') . \tag{10.2}$$

$\theta_\omega(r, r')$ is equal to one as long as $|\omega|$ is smaller than or at most as big as the Debye frequencies corresponding to the positions $r$ and $r'$. $\theta_\omega(r, r')$ is equal to zero otherwise. If it can be shown (and it can) that $K_\omega(r, r')$ is Hermitian, also $\sum_\omega \theta_\omega(r, r') \, K_\omega(r, r')$ is Hermitian.

A few further remarks regarding Eqs. (10.1) and (10.2) seem in order. The quantitity of main interest from the point of view of physics is the transition temperature for a second kind transition in the absence or presence of a magnetic field of a certain geometry and strength or, vice versa, the critical magnetic field as a function of temperature. The temperature, $T$, appears explicitly in Eq. (10.1) and also enters into the variable $\omega$. The kernel, $K_\omega(r, r')$, also depends upon the magnetic field as will become clear from the explicit formulations to be given below. The question is, whether temperature and magnetic field strength may be chosen in such a way that Eq. (10.1) admits a non-trivial solution. But not every such solution is of physical relevance. E.g., only the highest transition temperature obtained in this way for a given magnetic field will actually correspond to a transition.

The kernel, $K_\omega(r, r')$, is determined by the physics of the conduction electrons in their normal state. It may be expressed in terms of quantum mechanical Green's functions; cf. Eq. (2.16). In the method of the correlation function it is, however, claimed that the kernel may also be calculated in an essentially classical way. This method was developed heuristically in Chapter I. Microscopic proofs were supplied whenever possible. The method is only correct if several characteristic lengthes, including the linear dimensions of the sample, are sufficiently big as compared to the Fermi wavelength; cf. Eqs. (7.1), (7.22), and (7.54).

*b) Boltzmann Equation*

In this section the results of Chapter I shall be summarized for the case of a spherical Fermi surface (more accurately speaking: for spherical energy surfaces in the neighbourhood of the Fermi surface). The characteristic quantities in this case are the effective mass, $m^*$, the Fermi velocity, $v$, and their product, the Fermi momentum, $k_F$. These quantities enter into the density of states,

$$N = \frac{m^{*2} v}{2\pi^2}. \tag{10.3}$$

The academic case of free electrons is contained in this formulation if the effective mass, $m^*$, is replaced by the true electronic mass, $m$.

In Chapter I (Sections 4d, 5a, and 8) more general formulations were put forward and derived for Fermi surfaces of arbitrary shape provided that the conduction electrons occupy a single band. These more general formulations are in practice applicable only to single crystals since otherwise the crystalline structure of a metal, with varying orientation of the Fermi surface, would have to be taken into account. The statistical aspects of this structure and the boundaries of the crystallites pose difficult problems. It is, however, hoped that results obtained for spherical Fermi surfaces are essentially correct also for realistic metals with more general Fermi surfaces.

According to the method of correlation function, the kernel, $K_\omega(r, r')$, may be calculated from a distribution function for electrons (or carriers of information) which satisfy a Boltzmann equation (Section 5) with certain boundary conditions (Section 6) on surfaces and interfaces.
It is

$$\begin{aligned} K_\omega(r, r') &= m^{*2} v \oint g_\omega(r, v; r')\, d\Omega \\ &= 2\pi^2 N \oint g_\omega(r, v; r')\, d\Omega, \end{aligned} \tag{10.4}$$

where the density of states has been introduced with the help of Eq. (10.3). The integration is in Eq. (10.4) extended over all directions of the velocity vector $v$ with $|v| = v$ (Fermi velocity). The distribution function, $g_\omega(r, v; r')$, is defined by the (Laplace transformed) Boltzmann equation,

$$(2|\omega| + v \cdot \tilde{\partial} + nv\sigma) g_\omega(r, v; r')$$

$$- nv \oint \frac{d\sigma(\theta)}{d\Omega} g_\omega(r, v'; r') \, d\Omega' = \frac{1}{(2\pi)^2} \delta(r - r'), \quad (10.5)$$

plus certain boundary conditions to be formulated further below (Section 10c). Here, the symbol $\tilde{\partial}$ is the gauge invariant gradient for charge $-2e$ (electron pairs!),

$$\tilde{\partial} = \frac{\partial}{\partial r} + 2ieA(r), \quad (10.6)$$

where $A(r)$ is a vector potential of the external magnetic field. The two scattering terms contain the concentration (number per volume), $n$, of the impurity atoms as a factor. $v$ is the Fermi velocity already introduced.

$$\frac{d\sigma(\theta)}{d\Omega} \geqq 0 \quad (10.7)$$

is the differential scattering cross section for electrons impinging with Fermi velocity upon an impurity atom. $\theta$ denotes the scattering angle between the initial velocity, $v'$, and the final velocity, $v$; the modulus of both velocity vectors equals the Fermi velocity, $v$. The integration in Eq. (10.4) is extended over all directions of $v'$. The integrated scattering cross section, $\sigma$, is related to the differential scattering cross section by

$$\sigma = \oint \frac{d\sigma(\theta)}{d\Omega} \, d\Omega. \quad (10.8)$$

A more general formulation of the Boltzmann equation, with $d\sigma/d\Omega$ depending not only upon the scattering angle, was given in Eq. (7.68). In such a case, differential and integrated scattering cross sections are related by Eq. (5.10) and have to satisfy certain general conditions [Eq. (5.26)] following from the postulate of invariance under time reversal.

For $A(r) \equiv 0$ (no external magnetic field present) a physical interpretation is easily obtained if $g_\omega(r, v; r')$ is regarded as the Laplace transform,

$$g_\omega(r, v; r') = 2\pi \int_0^\infty h(r, v; r'; t) \exp\{-2|\omega| t\} \, dt, \quad (10.9)$$

of a time dependent distribution function, $h(r, v; r'; t)$. Evidently, $h(r, v; r'; t)$ satisfies the time dependent Boltzmann equation

$$\left(\frac{\partial}{\partial t} + v \cdot \frac{\partial}{\partial r} + nv\sigma\right) h(r, v; r'; t) - nv \oint \frac{d\sigma(\theta)}{d\Omega} h(r, v'; r'; t) \, d\Omega' = 0 \quad (10.10)$$

and the initial condition

$$h(r, v; r'; 0) = \frac{1}{(2\pi)^3} \delta(r - r'). \tag{10.11}$$

Eq. (10.10), in which the gradient, $\partial/\partial r$, occurs instead of the gauge invariant gradient [(Eq. (10.6))], is just the ordinary Boltzmann equation describing the propagation of electrons with position $r$ and velocity $v$ in the presence of statistically distributed scattering centres. Eq. (10.11) states that the electrons start at $t = 0$ from the position $r'$ with Fermi velocity and isotropic distribution of the velocity directions. Eq. (10.4) says that $K_\omega(r, r')$ is, apart from a factor, the Laplace transform of the corresponding density in real space. Eq. (10.1) may finally be interpreted by stating that these electrons propagate information as to the value of the pair potential from $r'$ to $r$.

This simple interpretation breaks down in the presence of a magnetic field when Eq. (10.5) holds. The gauge invariant gradient [Eq. (10.6)] would not appear in a Boltzmann equation proper. $g_\omega(r, v; r')$ is no longer a real function. It cannot be interpreted as a distribution function of particles. One might, however, still speak of a propagation of carriers of information which transport information as to the pair potential from $r'$ to $r$.

Eq. (10.5) is valid if the electrons are scattered by non-magnetic impurities. Some slight modifications are required in the presence of paramagnetic impurities. In this case it is

$$(2|\omega| + v \cdot \tilde{\partial} + n v \sigma_1) g_\omega(r, v; r')$$
$$- n v \oint \frac{d\sigma_2(\theta)}{d\Omega} g_\omega(r, v'; r') \, d\Omega' = \frac{1}{(2\pi)^2} \delta(r - r') \tag{10.12}$$

with

$$\sigma_1 \geqq 0, \quad \frac{d\sigma_2(\theta)}{d\Omega} \geqq 0. \tag{10.13}$$

If $\sigma_2$ is introduced by

$$\sigma_2 = \oint \frac{d\sigma_2(\theta)}{d\Omega} \, d\Omega, \tag{10.14}$$

$$\sigma_{sp} = \frac{\sigma_1 - \sigma_2}{2} \geqq 0 \tag{10.15}$$

is the spin-flip cross section.

The reader is reminded that $\hbar$, $c$, and $k_B$ have been put equal to 1. Magnetic field and vector potential are given in Gaussian units.

### c)  Boundary Conditions

The Boltzmann equations (10.5) or (10.12) have to be supplemented by boundary conditions. In a point $r$ on an external surface we postulate

$$g_\omega(r, v_{out}; r') = \int R(r; v_{out}, v_{in}) \, g_\omega(r, v_{in}; r') \, d\Omega_{in} \,. \qquad (10.16)$$

$r'$ is an arbitrary point in the sample, the starting point of the electrons or carriers of information in the interpretation given above. $v_{in}$ is the velocity vector of (incoming) electrons which approach the surface from the interior; $v_{out}$ belongs to (outgoing) electrons leaving the surface. The integration in Eq. (10.16) is extended over the semi-sphere of incoming directions. The modulus of both $v_{in}$ and $v_{out}$ equals the Fermi velocity, $v$.

The reflexion kernel, $R(r; v_{out}, v_{in})$, characterizes the reflexion properties of the surface in the point $r$. It has to obey some general conditions,

$$R(v_{out}, v_{in}) \gtreqless 0 \,, \qquad (10.17)$$

$$\int n \cdot v_{out} R(v_{out}, v_{in}) \, d\Omega_{out} = - n \cdot v_{in} \,, \qquad (10.18)$$

$$n \cdot v_{out} R(v_{out}, v_{in}) = - n \cdot v_{in} R(-v_{in}, -v_{out}) \,, \qquad (10.19)$$

$$\int R(v_{out}, v_{in}) \, d\Omega_{in} = 1 \,, \qquad (10.20)$$

which were derived in Section 6a. The position argument, $r$, has been dropped. $n$ is the normal unit vector in the surface point. If Eq. (10.19) is inserted into Eq. (10.18), Eq. (10.20) is obtained, and vice versa. It is, nevertheless, convenient to list and to apply all four conditions, Eq. (10.17) through (10.20). We mention a consequence of Eq. (10.18): the surface normal component of the particle flux vanishes,

$$\oint n \cdot v \, g_\omega(r, v; r') \, d\Omega = 0 \,. \qquad (10.21)$$

The kernel of specular reflexion is given by

$$R_{sp}(v_{out}, v_{in}) = \hat{\delta}(v_{out}, v_{in} - 2n(n \cdot v_{in})) \qquad (10.22)$$

with the delta function of directions, $\hat{\delta}(a, b)$, defined by Eq. (4.14). The kernel of diffuse reflexion is given by [12]

$$R_{diff}(v_{out}, v_{in}) = \frac{n \cdot v_{in}}{\pi v} \,. \qquad (10.23)$$

Here, $n$ is the surface normal pointing out of the sample. Usually not much is known about the detailed reflexion properties of a surface. In such a case it seems reasonable to work with the one-parametric family of reflexion kernels,

$$R(v_{out}, v_{in}) = (1 - p) \, R_{sp}(v_{out}, v_{in}) + p \, R_{diff}(v_{out}, v_{in}) \,, \qquad (10.24)$$

where the diffuseness, $p$, is bounded by

$$0 \leqq p \leqq 1 . \qquad (10.25)$$

It should be mentioned that *Reuter* and *Sondheimer* [27] used the complementary definition of $p$.

On an interface between different metals (say, metal 1 and metal 2) there are electrons (carriers of information) incoming from both metals and outgoing into both metals. There are now two reflexion kernels, $R_1(r; v_{1\,\text{out}}, v_{1\,\text{in}})$ and $R_2(r; v_{2\,\text{out}}, v_{2\,\text{in}})$, and two transmission kernels, $T_1(r; v_{1\,\text{out}}, v_{2\,\text{in}})$ and $T_2(r; v_{2\,\text{out}}, v_{1\,\text{in}})$, which enter into the boundary conditions,

$$\left.\begin{aligned}
g_\omega(r, v_{1\,\text{out}}; r') &= \int R_1(r; v_{1\,\text{out}}, v_{1\,\text{in}})\, g_\omega(r, v_{1\,\text{in}}; r')\, d\Omega_{1\,\text{in}} \\
&\quad + \int T_1(r; v_{1\,\text{out}}, v_{2\,\text{in}})\, g_\omega(r, v_{2\,\text{in}}; r')\, d\Omega_{2\,\text{in}}, \\
g_\omega(r, v_{2\,\text{out}}; r') &= \int R_2(r; v_{2\,\text{out}}, v_{2\,\text{in}})\, g_\omega(r, v_{2\,\text{in}}; r')\, d\Omega_{2\,\text{in}} \\
&\quad + \int T_2(r; v_{2\,\text{out}}, v_{1\,\text{in}})\, g_\omega(r, v_{1\,\text{in}}; r')\, d\Omega_{1\,\text{in}}.
\end{aligned}\right\} \quad (10.26)$$

Distribution functions containing a velocity vector with label 1 are boundary values of a solution to the Boltzmann equation for electrons in metal 1. The analogous statement is true for label 2.

The kernels have again to satisfy a number of general conditions,

$$R_1(v_{1\,\text{out}}, v_{1\,\text{in}}) \geqq 0, \quad T_1(v_{1\,\text{out}}, v_{2\,\text{in}}) \geqq 0 , \qquad (10.27)$$

$$N_1\{n \cdot v_{1\,\text{in}} + \int n \cdot v_{1\,\text{out}}\, R_1(v_{1\,\text{out}}, v_{1\,\text{in}})\, d\Omega_{1\,\text{out}}\}$$
$$= N_2 \int n \cdot v_{2\,\text{out}}\, T_2(v_{2\,\text{out}}, v_{1\,\text{in}})\, d\Omega_{2\,\text{out}}, \qquad (10.28)$$

$$n \cdot v_{1\,\text{out}}\, R_1(v_{1\,\text{out}}, v_{1\,\text{in}}) = -n \cdot v_{1\,\text{in}}\, R_1(-v_{1\,\text{in}}, -v_{1\,\text{out}}), \qquad (10.29)$$

$$N_1\, n \cdot v_{1\,\text{out}}\, T_1(v_{1\,\text{out}}, v_{2\,\text{in}}) = N_2\, n \cdot v_{2\,\text{in}}\, T_2(-v_{2\,\text{in}}, -v_{1\,\text{out}}), \qquad (10.30)$$

$$\int R_1(v_{1\,\text{out}}, v_{1\,\text{in}})\, d\Omega_{1\,\text{in}} + \int T_1(v_{1\,\text{out}}, v_{2\,\text{in}})\, d\Omega_{2\,\text{in}} = 1 . \qquad (10.31)$$

$N_1$ and $N_2$ are the densities of states in metal 1 and 2, respectively. There are further relations with labels 1 and 2 interchanged. The above conditions were derived in Section 6b.

It will be shown in Section 11 that Boltzmann equation plus boundary conditions define the distribution function, $g_\omega(r, v; r')$, uniquely. Eq. (10.4) then leads to a unique expression for the kernel, $K_\omega(r, r')$. A uniqueness theorem does of course not hold for Eq. (10.1). It was, however, pointed out above that only certain solutions to Eq. (10.1) are relevant from a physical point of view.

Some general properties of $K_\omega(r, r')$ which are true on the basis of the quantum mechanical definition (cf. Section 3) can be proved also from the Boltzmann equation formulation. This was shown in Section 5.

Further general properties (positiveness, boundedness) which apparently cannot be derived from quantum mechanics will be proved in Section 11. The classical orbits of the electrons or of the carriers of information will be exhibited in Section 12.

## 11. Uniqueness, Positiveness, and Boundedness of the Kernel. Transition Temperature

### a) Uniqueness

Boltzmann equation (10.5) and boundary conditions (10.16) and (10.26) together determine the distribution function, $g_\omega(r, v; r')$, uniquely [39] for $\omega \neq 0$. Therefore, the kernel, $K_\omega(r, r')$, is defined uniquely within the scope of the method of correlation function. A similar uniqueness theorem may be proved on a quantum mechanical basis as shown in Section 1b in connection with Eq. (1.15).

The proof starts in the usual manner. Let $g_\omega^{(1)}(r, v; r')$ and $g_\omega^{(2)}(r, v; r')$ be two solutions to Boltzmann equation plus boundary conditions. Their difference,

$$f_\omega(r, v) = g_\omega^{(1)}(r, v; r') - g_\omega^{(2)}(r, v; r'), \tag{11.1}$$

satisfies a homogeneous Boltzmann equation,

$$(2|\omega| + v \cdot \tilde{\partial} + nv\sigma) f_\omega(r, v) - nv \oint \frac{d\sigma(\theta)}{d\Omega} f_\omega(r, v') d\Omega' = 0 \tag{11.2}$$

and the same boundary conditions as the function $g_\omega(r, v; r')$. It shall be shown that it is

$$f_\omega(r, v) \equiv 0. \tag{11.3}$$

If the sample consists of a single metal, Eq. (11.2) is multiplied by $f_\omega^*(r, v)$. To the relation obtained in this way the complex conjugated expression is added. The result is integrated over all directions of $v$ and, with respect to $r$, over the whole sample. If the derivative term is transformed into a surface integral, we obtain

$$\oint d\Omega \int d^3r \, (4|\omega| + 2nv\sigma) |f_\omega(r, v)|^2$$

$$- \oint d\Omega \, d\Omega' \int d^3r \, nv \frac{d\sigma(\theta)}{d\Omega} (f_\omega^*(r, v) f_\omega(r, v') + f_\omega^*(r, v') f_\omega(r, v)) \tag{11.4}$$

$$+ \oint d\Omega \oint d^2r \cdot v |f_\omega(r, v)|^2 = 0.$$

In the surface integral, $d^2r$ is the vectorial surface element pointing to the exterior of the sample.

The scattering terms vanish if the metal does not contain impurity atoms ($n = 0$). Otherwise it is

$$2\sigma \oint d\Omega |f_\omega(r, v)|^2$$

$$- \oint d\Omega \, d\Omega' \, \frac{d\sigma(\theta)}{d\Omega} \left( f_\omega^*(r, v) \, f_\omega(r, v') + f_\omega^*(r, v') \, f_\omega(r, v) \right) \geq 0, \tag{11.5}$$

as shall be shown below. In an infinite sample the surface terms vanish provided that the distribution function goes to zero sufficiently strongly at infinity. Otherwise it is

$$\oint d\Omega \, \boldsymbol{n} \cdot \boldsymbol{v} |f_\omega(r, v)|^2 \geq 0, \tag{11.6}$$

as shall also be shown below. Here, $\boldsymbol{n}$ is the normal unit vector in the surface point $r$ which points to the exterior of the sample. From Eqs. (11.4), (11.5), and (11.6), we conclude

$$4 |\omega| \oint d\Omega \int d^3 r \, |f_\omega(r, v)|^2 \leq 0. \tag{11.7}$$

Eq. (11.3) follows for $\omega \neq 0$. It might be mentioned that a term with $\omega = 0$ does not occur on the righthand side of the gap equation (10.1).

The proof of Eq. (11.5) starts from

$$\oint d\Omega \, d\Omega' \, \frac{d\sigma(\theta)}{d\Omega} \, |f_\omega(r, v) - f_\omega(r, v')|^2 \geq 0, \tag{11.8}$$

which is correct because of Eq. (10.7). Writing down the various contributions to the squared modulus and making use of Eq. (10.8), we immediately obtain Eq. (11.5). Eq. (11.5) is, as it stands, applicable to non-magnetic impurities only. But also the analogous relation for paramagnetic impurities,

$$2\sigma_1 \oint d\Omega |f_\omega(r, v)|^2$$

$$- \oint d\Omega \, d\Omega' \, \frac{d\sigma_2(\theta)}{d\Omega} \left( f_\omega^*(r, v) \, f_\omega(r, r') + f_\omega^*(r, v') \, f_\omega(r, v) \right) \geq 0 \tag{11.9}$$

is easily proved. $\sigma_1$ and $d\sigma_2(\theta)/d\Omega$ are the scattering cross sections entering into Eq. (10.12). Eq. (11.9) is correct because of the inequality sign in Eq. (10.15).

If the differential scattering cross section depends not only upon the scattering angle, $\theta$, but separately upon the vectors $v$ and $v'$, use has also to be made of the invariance under time reversal [Eq. (5.26)]. It follows from this equation that not only Eq. (5.10) but also Eq. (5.32) holds. Also in the case of non-spherical Fermi surfaces in which the Boltzmann equation is given by Eq. (5.27), a recourse to the invariance under time reversal has to be made.

For the proof of Eq. (11.6), we start from

$$-\int d\Omega_{out} d\Omega_{in} \, \boldsymbol{n} \cdot \boldsymbol{v}_{out} R(\boldsymbol{r}; \boldsymbol{v}_{out}, \boldsymbol{v}_{in}) |f_\omega(\boldsymbol{r}, \boldsymbol{v}_{out}) - f_\omega(\boldsymbol{r}, \boldsymbol{v}_{in})|^2 \geqq 0. \qquad (11.10)$$

This inequality is true as a consequence of Eq. (10.17), since $\boldsymbol{n}$ points to the exterior and $\boldsymbol{v}_{out}$ to the interior of the sample. Various terms occur in the integrand on the lefthand side after the squared modulus has been evaluated. It is

$$\int d\Omega_{out} d\Omega_{in} \, \boldsymbol{n} \cdot \boldsymbol{v}_{out} R(\boldsymbol{r}; \boldsymbol{v}_{out}, \boldsymbol{v}_{in}) |f_\omega(\boldsymbol{r}, \boldsymbol{v}_{in})|^2$$
$$= -\int d\Omega_{in} \, \boldsymbol{n} \cdot \boldsymbol{v}_{in} |f_\omega(\boldsymbol{r}, \boldsymbol{v}_{in})|^2 \qquad (11.11)$$

because of Eq. (10.18). Further it is

$$\int d\Omega_{out} d\Omega_{in} \, \boldsymbol{n} \cdot \boldsymbol{v}_{out} R(\boldsymbol{r}; \boldsymbol{v}_{out}, \boldsymbol{v}_{in}) |f_\omega(\boldsymbol{r}, \boldsymbol{v}_{out})|^2$$
$$= \int d\Omega_{out} \, \boldsymbol{n} \cdot \boldsymbol{v}_{out} |f_\omega(\boldsymbol{r}, \boldsymbol{v}_{out})|^2 \qquad (11.12)$$

because of Eq. (10.20). Finally it is

$$\int d\Omega_{out} d\Omega_{in} \, \boldsymbol{n} \cdot \boldsymbol{v}_{out} R(\boldsymbol{r}; \boldsymbol{v}_{out}, \boldsymbol{v}_{in}) f_\omega^*(\boldsymbol{r}, \boldsymbol{v}_{out}) f_\omega(\boldsymbol{r}, \boldsymbol{v}_{in})$$
$$= \int d\Omega_{out} \, \boldsymbol{n} \cdot \boldsymbol{v}_{out} |f_\omega(\boldsymbol{r}, \boldsymbol{v}_{out})|^2 \qquad (11.13)$$

as a consequence of the boundary condition (10.16) which also holds for $f_\omega(\boldsymbol{r}, \boldsymbol{v})$. Making use also of the complex conjugate of Eq. (11.13) we finally arrive at Eq. (11.7). It seems noteworthy that almost the whole machinery of general conditions to be satisfied by the reflexion kernel enters into the proof of Eq. (11.7).

If a sample consists of several metals in electric contact, the contribution from each metal ($j$) has to be multiplied by the corresponding density of states, $N_j(>0)$, so that Eq. (11.4) is replaced by

$$\sum_j N_j \left[ \oint d\Omega \int_{V_j} d^3 r (4|\omega| + 2n_j v_j \sigma_j) |f_\omega(\boldsymbol{r}, \boldsymbol{v}_j)|^2 \right.$$

$$- \oint d\Omega d\Omega' \int_{V_j} d^3 r \, n_j v_j \frac{d\sigma_j(\theta)}{d\Omega} (f_\omega^*(\boldsymbol{r}, \boldsymbol{v}_j) f_\omega(\boldsymbol{r}, \boldsymbol{v}_j') + f_\omega^*(\boldsymbol{r}, \boldsymbol{v}_j') f_\omega(\boldsymbol{r}, \boldsymbol{v}_j)) \qquad (11.14)$$

$$\left. + \oint d\Omega \oint_{B_j} d^2 r \cdot \boldsymbol{v}_j |f_\omega(\boldsymbol{r}, \boldsymbol{v}_j)|^2 \right] = 0.$$

The volume integrals have to be done over the volume, $V_j$, of each particular metal. The factors in the scattering terms do in general vary from metal to metal; they may even be position dependent inside a particular metal. Eq. (11.5) is still valid for these scattering terms.

The boundary integrals with integration over the complete boundary, $B_j$, of each metal contain both surface and interface contributions. $d^2 r$ is the surface element pointing to the exterior of each particular metal. Eq. (11.6) is valid in surface points. In an interface point, there are contributions from both adjoining metals. It follows from the boundary conditions (10.26) and the general properties of the reflexion and transmission kernels that also the interface contributions to Eq. (11.14) are never negative. The proof proceeds in the same way as above for external surfaces. It is somewhat lengthy and shall not be reproduced here. We finally arrive at

$$4\,|\omega|\sum_j N_j \oint d\Omega \int_{V_j} d^3 r\,|f_\omega(r, v_j)|^2 \leqq 0 .$$ (11.15)

Since the density of states, $N_j$, is a positive quantity, Eq. (11.3) is again obtained for $\omega \neq 0$.

### b) Positiveness and Boundedness

The kernel is Hermitian,

$$K_\omega(r, r') = K_\omega^*(r', r) .$$ (11.16)

This follows immediately from its quantum mechanical definition (2.16) and can also be proved with the help of the method of the correlation function; cf. Section 5 b. It is sometimes convenient to regard the kernel as the matrix element of a Hermitian operator, $\mathbb{K}_\omega$, by postulating

$$K_\omega(r, r') = \langle r|\mathbb{K}_\omega| r'\rangle .$$ (11.17)

It is

$$\mathbb{K}_\omega \varDelta(r) = \int \langle r|\mathbb{K}_\omega| r'\rangle \varDelta(r')\, d^3 r' = \int K_\omega(r, r')\, \varDelta(r')\, d^3 r' .$$ (11.18)

We want to prove that it is [39]

$$0 < \mathbb{K}_\omega \leqq \frac{\pi N}{|\omega|} ,$$ (11.19)

if a sample consists of a single metal with density of states $N$. This is a shorthand notation for the assertion

$$0 < (\varDelta, \mathbb{K}_\omega \varDelta) \leqq \frac{\pi N}{|\omega|}\, (\varDelta, \varDelta)$$ (11.20)

with arbitrary $\varDelta(r) \not\equiv 0$. In Eq. (11.20), it is

$$(\varDelta, \varDelta) = \int |\varDelta(r)|^2\, d^3 r, \quad (\varDelta, \mathbb{K}_\omega \varDelta) = \int \varDelta^*(r)\, K_\omega(r, r')\, \varDelta(r')\, d^3 r\, d^3 r' .$$ (11.21)

$(\varDelta, \mathbb{K}_\omega \varDelta)$ is real since $\mathbb{K}_\omega$ is a Hermitian operator. A generalization of Eq. (11.19) to samples consisting of several metals will be formulated and proved further below.

In order to prove Eqs. (11.19) and (11.20), we introduce the distribution function

$$\varDelta_\omega(r, v) = \int g_\omega(r, v; r')\, \varDelta(r')\, d^3 r' ,$$ (11.22)

which is the unique solution to the Boltzmann equation

$$(2|\omega| + v \cdot \tilde{\partial} + nv\sigma) \Delta_\omega(r, v)$$

$$- nv \oint \frac{d\sigma(\theta)}{d\Omega} \Delta_\omega(r, v') \, d\Omega' = \frac{1}{(2\pi)^2} \Delta(r) \qquad (11.23)$$

with the appropriate boundary conditions. It is

$$\mathbb{K}_\omega \Delta(r) = 2\pi^2 N \oint \Delta_\omega(r, v) \, d\Omega \qquad (11.24)$$

as a consequence of Eqs. (11.18), (10.4), and (11.22). If the sample consists of a single metal, Eq. (11.23) is now multiplied by $\Delta_\omega^*(r, v)$. The complex conjugated relation is added. The sum is integrated over all directions of $v$ and over the whole sample. The same procedure as used in the uniqueness proof (Section 11 a) then leads to

$$4|\omega| N \oint d\Omega \int d^3r |\Delta_\omega(r, v)|^2 \leq \frac{1}{(2\pi^2)^2} (\Delta, \mathbb{K}_\omega \Delta), \qquad (11.25)$$

where use has also been made of Eq. (11.24) and of the Hermiticity of the kernel. Since the lefthand side is positive for $\Delta_\omega(r, v) \not\equiv 0$ we immediately arrive at

$$\mathbb{K}_\omega > 0. \qquad (11.26)$$

This is the first part of Eq. (11.19).

In order to also prove the second part, we first conclude

$$|\oint \Delta_\omega(r, v) \, d\Omega|^2 \leq 4\pi \oint |\Delta_\omega(r, v)|^2 \, d\Omega \qquad (11.27)$$

from

$$\oint |\Delta_\omega(r, v) - \Delta_\omega(r, v')|^2 \, d\Omega \, d\Omega' \geq 0. \qquad (11.28)$$

Using Eq. (11.27) together with Eq. (11.24), we obtain

$$\frac{|\omega|}{\pi N} (\Delta, \mathbb{K}_\omega^2 \Delta) \leq (\Delta, \mathbb{K}_\omega \Delta) \qquad (11.29)$$

from Eq. (11.25). With the help of Cauchy's inequality,

$$(\Delta, \mathbb{K}_\omega \Delta)^2 \leq (\Delta, \mathbb{K}_\omega^2 \Delta)(\Delta, \Delta), \qquad (11.30)$$

we finally derive

$$(\Delta, \mathbb{K}_\omega \Delta) \leq \frac{\pi N}{|\omega|} (\Delta, \Delta) \qquad (11.31)$$

or

$$\mathbb{K}_\omega \leq \frac{\pi N}{|\omega|} \qquad (11.32)$$

from Eq. (11.29). This completes the proof of Eq. (11.19) for a sample consisting of a single metal.

If the sample consists of several metals in electric contact, the assertion (11.19) is replaced by

$$0 < \mathbb{K}_\omega \leq \frac{\pi}{|\omega|} \mathbb{N},$$ (11.33)

where $\mathbb{N}$ is a positive operator with the matrix element

$$\langle r|\mathbb{N}|r'\rangle = N(r)\,\delta(r-r').$$ (11.34)

$N(r)$ is the density of states of the particular metal; it was called $N_j$ at several other occasions. Eq. (11.25) now reads

$$4|\omega|\sum_j N_j \oint d\Omega \int_{V_j} d^3r\,|\Delta_\omega(r, v_j)|^2 \leq \frac{1}{(2\pi^2)^2}(\Delta, \mathbb{K}_\omega \Delta).$$ (11.35)

Eq. (11.26) is again an immediate consequence. On the other hand, the same manipulations as made above lead to

$$\frac{|\omega|}{\pi}\left(\Delta, \mathbb{K}_\omega \frac{1}{\mathbb{N}} \mathbb{K}_\omega \Delta\right) \leq (\Delta, \mathbb{K}_\omega \Delta).$$ (11.36)

This relation replaces inequality (11.29). If the Cauchy inequality is used in the form

$$(\Delta, \mathbb{K}_\omega \Delta)^2 \leq \left(\Delta, \mathbb{K}_\omega \frac{1}{\mathbb{N}} \mathbb{K}_\omega \Delta\right)(\Delta, \mathbb{N}\Delta),$$ (11.37)

Eq. (11.33) is indeed obtained.

## c) Transition Temperature of a Single Metal

We regard a sample consisting of a single metal so that the coupling constant, $g$, the density of states, $N$, and the Debye frequency, $\omega_D$, are position independent. At a second kind transition, Eq. (10.1) (with the dash on the summation sign reminding of the cut-off) admits a non-trivial solution. It follows from this equation that it also is

$$(\Delta, \Delta) = g\,T\sum_\omega{}' (\Delta, \mathbb{K}_\omega \Delta).$$ (11.38)

The first inequality sign in Eq. (11.20) leads to the conclusion that Eq. (11.38) can be satisfied only for $g > 0$ (attractive electron-electron interaction).

If use is made also of the second sign of inequality in Eq. (11.20), we obtain

$$1 \leq g\,N\,\pi\,T\sum_\omega{}' \frac{1}{|\omega|}.$$ (11.39)

Since the $\omega$ summation is extended over the odd multiples of $\pi T$, Eq. (11.39) leads to

$$1 \leqq g N 2 \sum_{l=0}^{L} \frac{1}{2l+1} \,, \tag{11.40}$$

where $L$ is the biggest integer equal to or smaller than $(\omega_D/2\pi T) - \frac{1}{2}$. It is seen that the temperature, $T$, enters into Eq. (11.40) only through the value of $L$. The sum is easily evaluated in the so-called weak-coupling limit,

$$L \gg 1 . \tag{11.41}$$

It is

$$2 \sum_{l=0}^{L} \frac{1}{2l+1} = \ln(4\gamma L) + O\left(\frac{1}{L}\right) \tag{11.42}$$

with Euler's constant

$$\gamma = 1.781 \dots . \tag{11.43}$$

If we now put

$$2 \sum_{l=0}^{L} \frac{1}{2l+1} = \ln\left(\frac{2\gamma\,\omega_D}{\pi T}\right), \tag{11.44}$$

there occurs no longer the unphysical discrete cut-off.

If a temperature, $T_c$, is defined by

$$1 = g N \pi T_c \sum' \frac{1}{|\omega_c|} \tag{11.45}$$

(with $\omega_c$ equal to the odd multiples of $\pi T_c$) or

$$T_c = \frac{2\gamma}{\pi} \omega_D \exp\left(-\frac{1}{gN}\right) \tag{11.46}$$

with

$$\frac{2\gamma}{\pi} = 1.134 \dots , \tag{11.47}$$

we find from Eq. (11.39) that it always is

$$T \leq T_c . \tag{11.48}$$

There is certainly no second kind transition for temperatures bigger than $T_c$. We want to show that $T_c$ as defined by Eq. (11.46) actually is the transition temperature in the absence of a magnetic field and of paramagnetic impurities [22].

A direct proof of this statement is obtained by putting the pair potential equal to a constant,

$$\Delta(r) =: \Delta , \tag{11.49}$$

in Eq. (11.23). This Boltzmann equation is, for $A(r) \equiv 0$, solved by a distribution function $\Delta_\omega(r, v)$ which depends neither upon $r$ nor upon $v$. Actually, it is

$$\Delta_\omega(r, v) = \frac{1}{2|\omega|(2\pi)^2} \Delta ,   \tag{11.50}$$

since the scattering terms cancel each other as a consequence of Eq. (10.8). Also the boundary condition (10.16) is satisfied because of Eq. (10.20). Actually, the present solution is closely related to the microcanonical ensemble (5.3). Eq. (11.50) is the unique solution for the pair potential (11.49). If use is made of Eq. (11.24), Eq. (10.1) leads to Eq. (11.40) with a sign of equality.

An alternative proof makes use of the symmetry [Eq. (3.6)] and the sum rule (3.13), which are both valid for $A(r) \equiv 0$. It is directly seen that Eq. (10.1) is solved by a constant pair potential. Eq. (11.40) with equality sign is immediately obtained.

The weak coupling condition (11.41) may be stated in the form

$$2\pi T_c \ll \omega_D ,   \tag{11.51}$$

where the righthand side may also be replaced by the Debye temperature. Because of Eq. (11.46) it may also be formulated as

$$gN \ll 1 ,   \tag{11.52}$$

which literally means weak coupling.

According to the derivation given above, the transition temperature in the absence of a magnetic field does not depend upon the shape and size of the sample or upon detailed properties of the surface (reflexion kernel). But it is tacitly assumed that all linear dimensions are big as compared to the lattice constant (Fermi wavelength) since otherwise the method of the correlation function would not be applicable. The transition temperature does, according to the derivation, also not depend upon the concentration of non-magnetic impurities. Experimentally there is a slight concentration dependence which, however, is much weaker than in the case of paramagnetic impurities. It is seen from this discrepancy that it actually is an oversimplification to represent impurities by scattering centres.

The transition temperature is, in the presence of a magnetic field, always smaller than $T_c$. Indeed, if the transition would occur at $T_c$, Eq. (11.31) and so Eq. (11.28) would have to hold with the equality sign. Therefore,

$$\Delta_\omega(r, v) = : \frac{1}{4\pi} \Delta_\omega(r)   \tag{11.53}$$

would not depend upon the direction of $v$. The boundary conditions on the surface are satisfied because of Eq. (10.20). In the Boltzmann equation (11.23) the scattering terms cancel each other. It then follows from this equation that it is

$$\Delta_\omega(r) = \frac{1}{2\pi|\omega|}\Delta(r) \tag{11.54}$$

and

$$\tilde{\partial}\Delta_\omega(r) = 0. \tag{11.55}$$

Eq. (11.54) means that the transition indeed occurs at $T_c$. Combining Eqs. (11.54) and (11.55) and making use of the commutation relations (11.57) of the Cartesian components of $\tilde{\partial}$ we would obtain

$$H(r)\Delta(r) = 0. \tag{11.56}$$

If the magnetic field vanishes nowhere in the sample, Eq. (11.56) leads to $\Delta(r) \equiv 0$ which means that no transition occurs at $T = T_c$. If there are regions in which the magnetic field vanishes, the modulus of $\Delta(r)$ is constant in these regions as a consequence of Eqs. (11.54) and (11.55). If this constant were different from zero, Eq. (11.55) would not be satisfied at the boundary between magnetic field and field free region.

The commutation relations referred to above read

$$\tilde{\partial}_j\tilde{\partial}_k - \tilde{\partial}_k\tilde{\partial}_j = 2ie\,\varepsilon_{jkl}H_l(r). \tag{11.57}$$

The latin subscripts denote Cartesian components. $\varepsilon_{jkl}$ is the totally antisymmetric Ricci tensor with $\varepsilon_{123} = +1$. Summation over Cartesian subscripts appearing twice is tacitly assumed. Eq. (1.24) was used for the derivation of Eq. (11.57).

The transition temperature of a metal with paramagnetic impurities is easily calculated if the impurity concentration is homogeneous. Eq. (10.12) leads to

$$(2|\omega| + v\cdot\tilde{\partial} + nv\sigma_1)\,\Delta_\omega(r,v)$$

$$- nv\oint\frac{d\sigma_2(\theta)}{d\Omega}\,\Delta_\omega(r,v')\,d\Omega' = \frac{1}{(2\pi)^2}\Delta(r), \tag{11.58}$$

which replaces Eq. (11.23). Putting the vector potential identically equal to zero and the pair potential equal to a constant [Eq. (11.49)], we obtain the solution

$$\Delta_\omega(r,v) = \frac{1}{2(|\omega| + nv\sigma_{sp})(2\pi)^2}\Delta \tag{11.59}$$

with the spin-flip cross section, $\sigma_{sp}$, given by Eq. (10.15). Eq. (11.45) is replaced by [36]

$$1 = gN\pi T\sum'\frac{1}{|\omega| + nv\sigma_{sp}}, \tag{11.60}$$

where $T$ is now the transition temperature in the presence of paramagnetic impurities. The same result is easily derived with the help of the modified sum rule (9.29). The further evaluation of Eq. (11.60) is postponed to Section 14c. It will turn out that paramagnetic impurities lower the transition temperature as had to be expected because of Eq. (11.48).

## 12. Classical Orbits

The Boltzmann equation (10.5) describes the propagation of electrons or, in the presence of a magnetic field, of carriers of information. In this section, the classical orbits of the electrons or of the carriers of information shall be exhibited. Some of the results to be derived are useful for practical calculations. We hope that the discussion of orbits will also lead to a deeper physical understanding of the method of the correlation function.

### a) Clean Metal

In the absence of impurities (i.e., in a clean metal), the Boltzmann equation (10.5) reads

$$(2|\omega| + v \cdot \tilde{\partial}) \, g_\omega(r, v; r') = \frac{1}{(2\pi)^2} \, \delta(r - r') \, . \tag{12.1}$$

This equation shall first be solved in an infinitely extended sample and in the absence of a magnetic field, so that $\tilde{\partial}$ is replaced by the gradient, $\partial/\partial r$. The solution vanishing at infinity is then given by

$$g_\omega(r, v; r') = \frac{1}{(2\pi|r - r'|)^2 v} \, \exp\left[-\frac{2|\omega| \, |r - r'|}{v}\right] \hat{\delta}(r - r', v) \, . \tag{12.2}$$

The delta function of directions, $\hat{\delta}(a, b)$, was defined in Eq. (4.14). That Eq. (12.2) indeed solves Eq. (12.1) for $A(r) \equiv 0$ was shown in connection with Eq. (7.3). The arguments shall not be repeated here. According to the uniqueness theorem (Section 11a), Eq. (12.2) gives the unique solution for $\omega \neq 0$.

The general solution to Eq. (12.1) for $A(r) \equiv 0$ is obtained, if the general solution to the homogeneous equation is added on the righthand side of Eq. (12.2). The latter solution is given by $\exp\left[-\dfrac{2|\omega|}{v^2} v \cdot r\right]$ multiplied by an arbitrary function of $v$ and $v \times r$. It is instructive to see how the uniqueness theorem operates: the general solution to the homogeneous equation is excluded since it does not vanish and is not even bounded at infinity. This exclusion cannot be made for $\omega = 0$ for which case no uniqueness theorem exists.

The distribution function (12.2) is different from zero only on the straight line of direction $v$ starting in $r'$. It evidently represents a classical particle orbit and is, indeed, the Laplace transform, in the sense of Eq. (10.9), of a time dependent distribution function describing a propagation of particles starting with velocity $v$ in the position $r'$. Because of the exponential,

$$\exp\left[-\frac{2|\omega| \, |r - r'|}{v}\right] = \exp\left[-\frac{|r - r'|}{\xi_\omega}\right], \tag{12.3}$$

the distribution function decays or fades with a characteristic length

$$\xi_\omega = \frac{v}{2|\omega|}. \tag{12.4}$$

This decay is a direct mathematical consequence of the Laplace transformation mentioned.

Also the solution to Eq. (12.1) in the presence of a magnetic field is easily obtained. It is given by

$$g_\omega(r, v; r')$$

$$= \frac{1}{(2\pi|r-r'|)^2 v} \exp\left[-\frac{2|\omega|\,|r-r'|}{v} - 2ie\int_{r'}^{r} A(s)\cdot ds\right] \hat{\delta}(r-r', v) \tag{12.5}$$

and is (for $\omega \neq 0$) the only solution vanishing at infinity. The integral in the exponential is extended along the straight line from $r'$ to $r$. This solution is different from zero on the same straight line as the righthand side of Eq. (12.2). In this sense, it again represents a classical motion. There occurs also the same exponential decay as in Eq. (12.2). But in addition, there is a complex phase factor due to the vector potential. This phase factor would not be present in a distribution function of classical particles. It is of typically quantum-mechanical origin. We propose to speak of a propagation of carriers of information. We see that there is a complex phase factor connected with this propagation.

In a finite sample consisting of a single clean metal, the Boltzmann equation (12.1) is supplemented by the boundary condition (10.16) on the surface. The problem may be formulated by an equivalent integral equation,

$$g_\omega(r, v; r') = g_\omega^{(0)}(r, v; r')$$
$$- (2\pi)^2 \oint d^2 r'' \cdot v \int d\Omega' g_\omega^{(0)}(r, v; r'') R(r''; v, v') g_\omega(r, v'; r'), \tag{12.6}$$

where $g_\omega^{(0)}(r, v; r')$ is given by the righthand side of Eq. (12.5), if the straight line connecting the points $r'$ and $r$ does not cut the surface, and is put equal to zero, if it does. $r''$ is a point on the surface, $d^2 r''$ the vectorial surface element pointing towards the exterior. The $r''$ integration is extended over the complete surface. The reflexion kernel, $R(r''; v, v')$, is defined only for incoming velocities $v'$ and outgoing velocities $v$ in the sense explained in connection with Eq. (10.16). The $v'$ integration has, therefore, to be extended over the semisphere of incoming directions only.

Eq. (12.6) is easily confirmed. The Boltzmann equation (12.1) is satisfied in any point $r$ in the interior of the sample, since it is satisfied by $g_\omega^{(0)}(r, v; r')$. Only the first term on the righthand side of Eq. (12.6) contributes to the inhomogeneity in the Boltzmann equation. In order to also check the boundary condition (10.16), we let the point $r$ approach the surface from the interior. To the distribution function $g_\omega(r, v; r')$ with outgoing velocity vector $v$ then only the second term on the right-hand side of Eq. (12.6) contributes but not the first one. The $r''$ integration may be transformed into an integration with respect to the direction of the vector $r - r''$. It is

$$d^2 r'' \cdot v = -(r - r'')^2 v \, d\Omega . \qquad (12.7)$$

If $g_\omega^{(0)}(r, v; r'')$ is inserted from Eq. (12.5) with the exponential replaced by one because of the small distance between $r$ and $r''$, the $r''$ integration may be done with the help of Eq. (12.7). The result is

$$g_\omega(r, v; r') = \int R(r; v, v') \, g_\omega(r, v'; r') \, d\Omega' \qquad (12.8)$$

which is just the boundary condition required. $v$ is an outgoing, $v'$ an ingoing velocity vector.

A simple intuitive interpretation may be given to Eq. (12.6). The first term on the righthand side represents the particles (or carriers of information), which travel on a straight line from $r'$ to $r$ without hitting the surface. The integral represents particles which are reflected once or several times by the surface. They had their last collision with the surface in the point $r''$ before they arrive in $r$.

The classical orbits are even more clearly exhibited if Eq. (12.6) is solved in an iterative manner,

$$g_\omega(r, v; r') = g_\omega^{(0)}(r, v; r')$$
$$-(2\pi)^2 \oint d^2 r'' \cdot v \int d\Omega' g_\omega^{(0)}(r, v; r'') R(r''; v, v') g_\omega^{(0)}(r'', v'; r') + \cdots . \qquad (12.9)$$

The second term on the righthand side now represents particles which hit the surface exactly once before arriving at $r$. The higher order terms represent several surface reflexions. There are no higher order terms if the sample is a semi-infinite metal with a plane boundary.

It is instructive to calculate the kernel with the help of Eq. (10.4) and to insert $g_\omega^{(0)}(r, v; r')$ from Eq. (12.5). The result is

$$K_\omega(r, r') = \frac{N}{2v} \left\{ \frac{\theta(r, r')}{(r - r')^2} \exp\left[ -\frac{|r - r'|}{\xi_\omega} - 2ie \int_{r'}^{r} A(s) \cdot ds \right] \right.$$
$$- \oint d^2 r'' \cdot \hat{v} R(r''; v, v') \frac{\theta(r, r'') \theta(r'', r')}{|r - r''|^2 |r'' - r'|^2} \qquad (12.10)$$
$$\left. \cdot \exp\left[ -\frac{|r - r''| + |r'' - r'|}{\xi_\omega} - 2ie \int_{r' - r'' - r} A(s) \cdot ds \right] + \cdots \right\} .$$

The function $\theta(r, r')$ is equal to 1 if the straight line from $r'$ to $r$ does not cut the surface; it is equal to zero otherwise. The first term in the curly bracket represents the direct path. The integral is extended along the straight line from $r'$ to $r$. The second term represents those paths which hit the surface (in the point $r''$) exactly once. The total length of this path appears in the damping term. The integral is extended along the actual path, i.e., along a straight line from $r'$ to $r''$ and from there again along a straight line to $r$. The velocity vector $v$ is parallel to $r - r''$; $\hat{v}$ is the corresponding unit vector. The vector $v'$ is parallel to $r'' - r'$. Higher order terms again represent several surface reflexions. The Hermiticity of the kernel [Eq. (11.16)] is immediately proved for the terms written down in Eq. (12.10) if use is made of Eq. (10.19).

If a sample consists of several metals in electric contact, the same procedure can also be applied to the interfaces.

In any case the contributions from the various paths leading from the point $r'$ to the point $r$ are added to give $K_\omega(r, r')$. All contributions are real and non-negative in the absence of a magnetic field [$A(r) \equiv 0$]. The contributions from individual paths are multiplied by different phase factors but the modulus is unaltered if a magnetic field is present. Therefore, the modulus of $K_\omega(r, r')$ is, in the presence of a magnetic field, at most as big as $K_\omega(r, r')$ in the field free case.

### b) Metal with Impurities

The Boltzmann equation is now given by Eq. (10.5). Only infinitely extended metals shall be considered so that $g_\omega(r, v; r')$ has to vanish at infinity but no detailled boundary conditions have to be satisfied. Otherwise, the present considerations would have to be combined with those of Section 12a.

In order to again exhibit classical orbits we first introduce a function $g_\omega^{(\sigma)}(r, v; r')$ satisfying

$$(2|\omega| + v \cdot \tilde{\partial} + nv\sigma) g_\omega^{(\sigma)}(r, v; r') = \frac{1}{(2\pi)^2} \delta(r - r'). \qquad (12.11)$$

The solution vanishing at infinity is given by

$$g_\omega^{(\sigma)}(r, v; r') \qquad (12.12)$$

$$= \frac{1}{(2\pi|r - r'|)^2 v} \exp\left[ -\frac{|r - r'|}{\xi_\omega} - \int_{r'}^{r} n(s)\sigma|\mathrm{d}s| - 2ie\int_{r'}^{r} A(s) \cdot \mathrm{d}s \right] \hat{\delta}(r - r', v).$$

The integrations are extended along the straight classical orbit leading from $r'$ to $r$. If the concentration, $n$, of impurities is position independent, Eq. (12.12) may be written as

$$g_\omega^{(\sigma)}(r, v; r')$$

$$= \frac{1}{(2\pi|r - r'|)^2 v} \exp\left[ -\frac{|r - r'|}{\zeta_\omega} - 2ie\int_{r'}^{r} A(s) \cdot \mathrm{d}s \right] \hat{\delta}(r - r', v). \qquad (12.13)$$

The decay length, $\zeta_\omega$, is now given by

$$\frac{1}{\zeta_\omega} := \frac{1}{\xi_\omega} + \frac{1}{l} \tag{12.14}$$

where $l$ is the electronic mean free path,

$$l := \frac{1}{n\sigma} . \tag{12.15}$$

$\zeta_\omega$ approaches $\xi_\omega$ for $l \gg \xi_\omega$ and $l$ for $l \ll \xi_\omega$. $g_\omega^{(\sigma)}(r, v; r')$ is the (Laplace transformed) distribution function of particles (or carriers of information) traveling from $r'$ to $r$ without suffering a single scattering.

The Boltzmann equation (10.5) is now replaced by an equivalent integral equation,

$$g_\omega(r, v; r') = g_\omega^{(\sigma)}(r, v; r')$$

$$+ (2\pi)^2 \int d^3 r'' d\Omega' \, g_\omega^{(\sigma)}(r, v; r'') \, nv \frac{d\sigma(\theta)}{d\Omega} \, g_\omega(r'', v'; r') . \tag{12.16}$$

The first term on the righthand side represents particles which propagate from $r'$ to $r$ without being scattered by impurities. The second term represents those particles which underwent scatterings before reaching $r$ and actually had their last collision at $r''$. The integration with respect to $r''$ indicates that an averaging process with respect to the position of impurity atoms is performed. If the concentration, $n$, is position dependent it has to be taken at the position $r''$. $\theta$ is the scattering angle between the velocity vectors $v'$ and $v$. Eq. (12.16) is essentially identical with Eq. (7.67), which there occurred as an intermediate step in the microscopic proof of the method of the correlation function.

Eq. (12.16) corresponds to Eq. (12.6). An iterative solution to the integral equation may again be produced. The analogue to Eq. (12.10) is given by

$$K_\omega(r, r') = \frac{N}{2v} \left\{ \frac{1}{|r-r'|^2} \exp\left[ -\frac{|r-r'|}{\zeta_\omega} - 2ie \int_{r'}^{r} A(s) \cdot ds \right] \right.$$

$$+ \int d^3 r'' n \frac{d\sigma(\theta)}{d\Omega} \frac{1}{|r-r''|^2 |r''-r'|^2} \tag{12.17}$$

$$\left. \cdot \exp\left[ -\frac{|r-r''|+|r''-r'|}{\zeta_\omega} - 2ie \int_{r'-r''-r} A(s) \cdot ds \right] + \cdots \right\}.$$

Here, $\theta$ is the scattering angle between the directions of the particle orbit before and after the collision, i.e., between the vectors $r''-r'$ and $r-r''$.

The interpretation of the righthand side of Eq. (12.17) in terms of orbits of electrons (carriers of information) is straightforward. The various terms correspond to various numbers of impurity collisions (zero, one, ...). The contributions from the various zig-zag paths are added in order to obtain $K_\omega(r, r')$. All contributions are real and non-negative in the absence of a magnetic field $[A(r) \equiv 0]$. The contributions from individual paths are unaltered in modulus but multiplied by a quantum-mechanical phase factor in the presence of a magnetic field. The modulus of $K_\omega(r, r')$ is, therefore, smaller in the presence of a magnetic field than in the field free case.

## 13. Ginzburg-Landau Approximation

In 1950, *Landau* and *Ginzburg* [8] proposed a phenomenological theory of superconductivity. The new element in it was a kind of one-particle wave function. With the microscopic understanding of the phenomenon of superconductivity achieved in the BCS theory, the Ginzburg-Landau wave function can be identified with the wave function of condensated electron pairs or, what is essentially the same thing, with the pair potential, $\Delta(r)$. *Ginzburg* and *Landau* proposed a non-linear differential equation for their wave function. At a second kind transition, the pair potential goes continuously to zero, so that a linear differential equation is obtained.

The linearized Ginzburg-Landau (GL) equation has the simple form

$$\tilde{\partial} \cdot \tilde{\partial}\, \Delta(r) = - \lambda(T)\, \Delta(r)\,. \tag{13.1}$$

$\tilde{\partial}$ is the gauge invariant gradient defined in Eq. (10.6) (*Ginzburg* and *Landau* themselves did not imagine electron pairs and, therefore, omitted the factor of 2 in the gauge invariant gradient). The quantity $\lambda(T)$ is a given function of temperature. It is non-negative for $T \leqq T_c$ and vanishes for $T = T_c$. It is sometimes convenient to introduce a length, $\xi(T)$, by

$$\lambda(T) =: \frac{1}{\xi^2(T)}\,. \tag{13.2}$$

Eq. (13.1) has to be supplemented by boundary conditions which, however, shall be postponed to Section 15.

A microscopic derivation of the GL equation was given by *Gorkov* [40, 41] on the basis of his version of the BCS theory. *Gorkov's* derivation shall in this section essentially be reproduced on the basis of the method of the correlation function [7]. The summation of infinitely many graphs, which *Gorkov* had to perform for a metal containing impurities, was done once for ever in the proof of the Boltzmann equation (Sections 7 and 8).

It follows from the assumptions to be made in the derivation of the linearized GL equation that this equation is only valid for temperatures close to the transition temperature, $T_c$. It will turn out in Section 15 that additional requirements as to the linear dimensions of the sample occur for finite samples.

### a) Derivation of the Linearized Ginzburg-Landau Equation for Clean Metals

In the absence of impurities we start from

$$(2|\omega| + v \cdot \tilde{\partial}) \, \Delta_\omega(r, v) = \frac{1}{(2\pi)^2} \, \Delta(r) \, ; \qquad (13.3)$$

cf. Eq. (11.23). The distribution function $\Delta_\omega(r, v)$ was defined in Eq. (11.22). Regarding the derivative term as a small quantity, we may solve Eq. (13.3) in an iterative manner,

$$\Delta_\omega(r, v) = \frac{1}{(2\pi)^2} \sum_{l=0}^{\infty} \frac{(-v \cdot \tilde{\partial})^l}{(2|\omega|)^{l+1}} \, \Delta(r) \, . \qquad (13.4)$$

Only the first three terms on the righthand side shall be kept in order to produce a second order differential equation for $\Delta(r)$. Eq. (13.4) is now inserted into the gap equation (10.1) or into the equation

$$\Delta(r) = 2\pi^2 g N T \sum_\omega{}' \oint \Delta_\omega(r, v) \, d\Omega \, . \qquad (13.5)$$

Making use of

$$\oint d\Omega = 4\pi, \quad \oint v \, d\Omega = 0, \quad \oint (v \cdot a)(v \cdot b) \, d\Omega = \frac{4\pi}{3} \, v^2 a \cdot b \, , \qquad (13.6)$$

we obtain

$$\Delta(r) = 2\pi g N T \sum_\omega{}' \left( \frac{1}{2|\omega|} + \frac{v^2}{3(2|\omega|)^3} \, \tilde{\partial} \cdot \tilde{\partial} + \cdots \right) \Delta(r) \qquad (13.7)$$

and have so essentially derived the linearized GL equation (13.1). The dash on the summation sign shall remind of the cut-off to be made at $\omega = \pm \omega_D$.

In the absence of a magnetic field $[A(r) \equiv 0]$, the pair potential may be chosen constant in space. Eq. (13.7) then reduces to

$$1 = 2\pi g N T \sum_\omega{}' \frac{1}{2|\omega|} \, , \qquad (13.8)$$

which is identical with Eq. (11.45) determining the transition temperature, $T_c$. Unter the assumption of weak coupling [Eq. (11.51)], the first sum on the righthand side of Eq. (13.7) may be done in exactly the same way as in Section 11c. As a result, the transition temperature may be introduced by putting

$$2\pi g N T \sum_{\omega}' \frac{1}{2|\omega|} = 1 + g N \ln\left(\frac{T_c}{T}\right).$$  (13.9)

The cut-off may be neglected in the second sum on the righthand side of Eq. (13.7) which is strongly convergent. The sum may then be done with the help of Riemann's zeta function,

$$2\pi T \sum_{\omega} \frac{v^2}{3(2|\omega|)^3} = \frac{7\,\zeta(3)}{12} \xi_0^2(T).$$  (13.10)

Here it is*

$$\xi_0(T) := \frac{v}{2\pi T}$$  (13.11)

and

$$\zeta(3) := \sum_{l=1}^{\infty} \frac{1}{l^3} = 1.202 \ldots .$$  (13.12)

Combining now everything, we arrive at the linearized GL equation (13.1) with [40]

$$\lambda(T) = \frac{12 \ln\left(\dfrac{T_c}{T}\right)}{7\,\zeta(3)\,\xi_0^2(T)} \approx \frac{12}{7\,\zeta(3)\,\xi_0^2(T_c)}\,\frac{T_c - T}{T_c}.$$  (13.13)

The last expression holds in the limit $T \to T_c$. We shall see that the GL approximation is reasonable only in this limit.

Cutting off the expansion (13.4) can indeed only be justified if the pair potential, $\Delta(r)$, is a slowly varying function on the range of the kernel $\sum' K_\omega(r, r')$. It follows from Eqs. (12.5) and (13.11) that this range is roughly given by $\xi_0(T)$. On the other hand, the GL equation (13.1) defines the characteristic length $\xi(T)$ [Eq. (13.2)]. We have to postulate

$$\xi(T) \gg \xi_0(T)$$  (13.14)

in order that the GL approximation is valid. Because of Eq. (13.13), this means that $T$ has to be close to $T_c$. Since not the simple gradient but $\tilde{\partial}$ occurs in the GL equation, actually some sophistication of the argument is required.

The linearized GL equation replaces Boltzmann equation (13.3) and gap equation (10.1). Therefore, a second kind transition between normal

---

* A slightly different length, $\xi_0 = (\gamma/\pi^2)(v/T_c)$, was defined in the classical BCS paper. Since it is $\gamma/\pi^2 = 1.8 \ldots$ and $1/2\pi = 1.6 \ldots$, the BCS $\xi_0$ is close to our $\xi_0(T_c)$.

and superconducting state may occur whenever the linearized GL equation admits a non-trivial solution behaving properly at infinity or satisfying appropriate boundary conditions. That not every such solution actually corresponds to a second kind transition was already discussed in Section 10.

The derivation of the linearized GL equation may be generalized to single crystals with a Fermi surface of arbitrary shape [42].

### b) Derivation of the Linearized Ginzburg-Landau Equation for Metals Containing Impurities

The derivation of the linearized GL equation may be extended to metals containing impurities. In this case, the Boltzmann equation (11.23) holds. Again the derivative term shall be regarded as a small quantity. With the expansion

$$\Delta_\omega(r, v) = \sum_{l=0}^\infty \Delta_\omega^{(l)}(r, v), \qquad (13.15)$$

we, therefore, postulate

$$(2|\omega| + nv\sigma)\, \Delta_\omega^{(l)}(r, v) - nv \oint \frac{d\sigma(\theta)}{d\Omega}\, \Delta_\omega^{(l)}(r, v')\, d\Omega'$$

$$= \begin{cases} \dfrac{1}{(2\pi)^2}\, \Delta(r) & (l=0), \\[2mm] -v \cdot \tilde{\partial}\, \Delta_\omega^{(l-1)}(r, v) & (l \geq 1). \end{cases} \qquad (13.16)$$

The uniqueness proof of Section 11a may be extended to Eq. (13.16) and is actually simpler in this case: Eq. (13.16) admits a unique solution provided that the righthand side is given.

The first Eq. (13.16) is solved by a $\Delta_\omega^{(0)}(r, v)$ not depending upon the direction of $v$. It is

$$\Delta_\omega^{(0)}(r, v) = \frac{1}{(2\pi)^2\, 2|\omega|}\, \Delta(r) =: \Delta_\omega^{(0)}(r). \qquad (13.17)$$

The scattering terms in Eq. (13.16) compensate each other because of Eq. (10.8). Eq. (13.17) is identical with the non-derivative term in Eq. (13.4). In order to obtain $\Delta_\omega^{(1)}(r, v)$, we make the ansatz

$$\Delta_\omega^{(1)}(r, v) = -a\, v \cdot \tilde{\partial}\, \Delta_\omega^{(0)}(r), \qquad (13.18)$$

where the constant, $a$, remains to be determined. Since it is

$$\oint \frac{d\sigma(\theta)}{d\Omega}\, v'\, d\Omega' = v\sigma^{(1)} \qquad (13.19)$$

with

$$\sigma^{(1)} := \oint \frac{d\sigma(\theta)}{d\Omega} \cos\theta \, d\Omega, \tag{13.20}$$

we derive

$$\Delta_\omega^{(1)}(\boldsymbol{r}, \boldsymbol{v}) = - \frac{1}{2|\omega| + nv(\sigma - \sigma^{(1)})} \, \boldsymbol{v} \cdot \tilde{\partial} \, \Delta_\omega^{(0)}(\boldsymbol{r}). \tag{13.21}$$

The righthand side may be expressed in terms of the transport cross section,

$$\sigma_{\mathrm{tr}} := \sigma - \sigma^{(1)}. \tag{13.22}$$

For the derivation of a second order differential equation, we do not have to go beyond $\Delta_\omega^{(2)}(\boldsymbol{r}, \boldsymbol{v})$ and actually need this quantity only integrated with respect to $\boldsymbol{v}$ over all directions. It follows from Eq. (13.16) with $l = 2$ and from Eq. (10.8) that it is

$$2|\omega| \oint \tilde{\partial} \, \Delta_\omega^{(2)}(\boldsymbol{r}, \boldsymbol{v}) \, d\Omega = - \oint \boldsymbol{v} \cdot \tilde{\partial} \, \Delta_\omega^{(1)}(\boldsymbol{r}, \boldsymbol{v}) \, d\Omega. \tag{13.23}$$

Inserting Eq. (13.21) into the righthand side, we arrive at

$$\oint \Delta_\omega^{(2)}(\boldsymbol{r}, \boldsymbol{v}) \, d\Omega = \frac{1}{2|\omega| \, (2|\omega| + nv\sigma_{\mathrm{tr}})} \, \frac{4\pi v^2}{3} \, \tilde{\partial} \cdot \tilde{\partial} \, \Delta_\omega^{(0)}(\boldsymbol{r}), \tag{13.24}$$

where use was made of Eq. (13.6). The impurity concentration was assumed constant in space; cf., however, below.

Inserting Eq. (13.15), with terms up to $l = 2$ retained, into Eq. (13.5), we obtain

$$\Delta(\boldsymbol{r}) = 2\pi g N T \sum_\omega{}' \left( \frac{1}{2|\omega|} + \frac{v^2}{3(2|\omega|)^2 (2|\omega| + nv\sigma_{\mathrm{tr}})} \, \tilde{\partial} \cdot \tilde{\partial} + \cdots \right) \Delta(\boldsymbol{r}). \tag{13.25}$$

Eq. (13.9) may again be applied. The cut-off may be neglected in the second term on the righthand side. Again, the GL equation (13.1) is obtained with $\lambda(T)$ given by [41]

$$\lambda(T) = \frac{12}{7\zeta(3)\chi\left(\dfrac{\xi_0(T_c)}{l_{\mathrm{tr}}}\right) \xi_0^2(T_c)} \, \frac{T_c - T}{T_c}. \tag{13.26}$$

$l_{\mathrm{tr}}$ is the transport length defined by

$$l_{\mathrm{tr}} := \frac{1}{n\sigma_{\mathrm{tr}}}. \tag{13.27}$$

The function $\chi(\varrho)$ is given by

$$\chi(\varrho) := \frac{8}{7\zeta(3)} \sum_{l=0}^{\infty} \frac{1}{(2l+1)^2 (2l+1+\varrho)}. \tag{13.28}$$

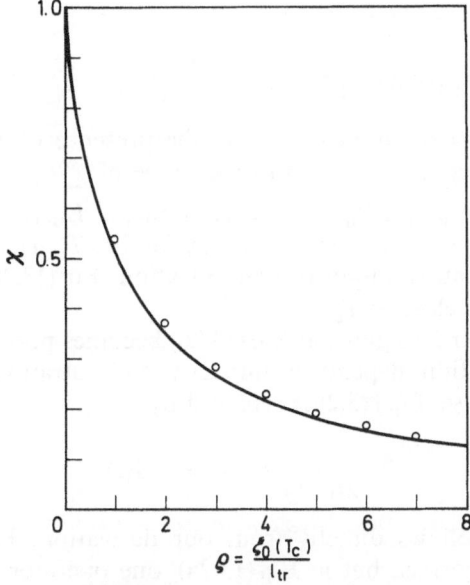

Fig. 1. The function $\chi(\varrho)$ entering into Eq. (13.26). The solid curve represents the function which is defined in Eq. (13.28). The circles represent the interpolation formula (13.32)

The function $\chi(\varrho)$ is shown in Fig. 1. It is a monotonically decreasing function of $\varrho$. Therefore, $\lambda(T)$ is enhanced by the alloying of impurities.

It is $\chi(\varrho) = 1$ in the clean limit $\varrho \ll 1$ $[l_{tr} \gg \xi_0(T_c)]$. It is

$$\chi(\varrho) = \frac{6\zeta(2)}{7\zeta(3)} \frac{1}{\varrho} \tag{13.29}$$

with

$$\zeta(2) := \sum_{l=1}^{\infty} \frac{1}{l^2} = \frac{\pi^2}{6} = 1.645\ldots \tag{13.30}$$

for $\varrho \gg 1$. This is the so-called dirty limit,

$$l_{tr} \ll \xi_0(T_c). \tag{13.31}$$

A useful interpolation between the two limiting cases is given by

$$\chi(\varrho) \approx \frac{1}{1 + \dfrac{7\zeta(3)}{6\zeta(2)}\varrho} = \frac{1}{1 + 0.853\varrho}. \tag{13.32}$$

This interpolation formula is represented by circles in Fig. 1. The function $\chi(\varrho)$ may be expressed by the digamma function,

$$\psi(z) = \frac{d \ln \Gamma(z)}{dz} = -\ln \gamma + \sum_{l=1}^{\infty} \left( \frac{1}{l} - \frac{1}{l+z-1} \right), \tag{13.33}$$

with $\gamma$ given by Eq. (11.43). It is

$$\sum_{l=0}^{\infty} \frac{1}{(2l+1)^2 (2l+1+\varrho)} = \frac{\pi^2}{8\varrho} - \frac{1}{2\varrho^2} \left[ \psi\left(\frac{1}{2} + \frac{\varrho}{2}\right) - \psi\left(\frac{1}{2}\right) \right]. \qquad (13.34)$$

The GL approximation is, also in the presence of impurities, only valid if $\xi(T)$ is big as compared to the range of $\sum' K_\omega(\mathbf{r}, \mathbf{r}')$. This range is of the order of $\sqrt{\xi_0(T)l_{\mathrm{tr}}}$ in the dirty limit; cf. Eq. (14.13). At least in this limit, the approximation is valid only for $T \approx T_c$. This is presumably true also for arbitray impurity concentration. Eq. (13.26) was already formulated for $T$ close to $T_c$.

The parameter $\lambda$ as given in Eq. (13.26) becomes position dependent, $\lambda(\mathbf{r}, T)$, for position dependent impurity concentration or transport length. In this case, Eq. (13.2) is replaced by

$$\tilde{\partial} \cdot \frac{1}{\lambda(\mathbf{r}, T)} \tilde{\partial} \Delta(\mathbf{r}) = -\Delta(\mathbf{r}). \qquad (13.35)$$

This assertion follows directly from our derivation: Eqs. (13.23) and (13.21) are still correct, but in Eq. (13.24), one operator $\tilde{\partial}$ has to stand in front of the whole expression.

In the presence of paramagnetic impurities, the derivation is based upon Eq. (10.12) or, rather, upon the corresponding equation for $\Delta_\omega(\mathbf{r}, \mathbf{v})$. Eq. (13.25) is replaced by

$$\Delta(\mathbf{r}) = 2\pi g N T$$

$$(13.36)$$

$$\cdot \sum_\omega' \left( \frac{1}{2(|\omega| + nv\sigma_{\mathrm{sp}})} + \frac{v^2}{12(|\omega| + nv\sigma_{\mathrm{sp}})^2 (2|\omega| + nv\sigma_{\mathrm{tr}})} \partial \cdot \partial + \cdots \right) \Delta(\mathbf{r}),$$

with the spin-flip cross section, $\sigma_{\mathrm{sp}}$, given by Eq. (10.15) and

$$\sigma_{\mathrm{tr}} := \sigma_1 - \int \frac{d\sigma_2(\theta)}{d\Omega} \cos\theta \, d\Omega. \qquad (13.37)$$

More compact forms of the GL equation may be derived in various ways [36, 4] from Eq. (13.36).

### c) Generalized Ginzburg-Landau Equation

Derivatives of higher than the second order shall now be retained. This is very simple for clean metals. In this case, all higher order terms are already contained in Eq. (13.4).

In the presence of impurities, it is advisable to first calculate $\Delta_\omega^{(2)}(\mathbf{r}, \mathbf{v})$ explicitly from Eq. (13.16). Having inserted Eq. (13.21) on the righthand side, we make the ansatz

$$\Delta_\omega^{(2)}(\mathbf{r}, \mathbf{v}) = (b v_j v_k + c v^2 \delta_{jk}) \frac{1}{2|\omega| + n v \sigma_{tr}} \tilde{\partial}_j \tilde{\partial}_k \Delta_\omega^{(0)}(\mathbf{r}) \qquad (13.38)$$

with constants $b$ and $c$ to be determined. Summation over Cartesian subscripts appearing twice is tacitly assumed. It is

$$b = \frac{1}{2|\omega| + \frac{3}{2} n v (\sigma - \sigma^{(2)})}, \qquad c = \frac{n v (\sigma - \sigma^{(2)})}{4|\omega|} b \qquad (13.39)$$

with

$$\sigma^{(2)} := \oint \frac{d\sigma(\theta)}{d\Omega} \cos^2 \theta \, d\Omega. \qquad (13.40)$$

In the derivation, use was made of

$$\oint \frac{d\sigma(\theta)}{d\Omega} v_j' v_k' \, d\Omega' = \frac{1}{2} (3\sigma^{(2)} - \sigma) v_j v_k + \frac{1}{2} (\sigma - \sigma^{(2)}) v^2 \delta_{jk} \qquad (13.41)$$

and of the fact that the tensors $v_j v_k$ and $\delta_{jk}$ are linearly independent. Eq. (13.24) is rederived with the help of Eq. (13.6) since it is

$$\frac{b}{3} + c = \frac{1}{6|\omega|}. \qquad (13.42)$$

Eq. (13.41) is proved by putting

$$\oint \frac{d\sigma(\theta)}{d\Omega} v_j' v_k' \, d\Omega' = \delta v_j v_k + \varepsilon v^2 \delta_{jk}, \qquad (13.43)$$

which follows from rotational invariance. In order to derive the values of the constants, $\delta$ and $\varepsilon$, we first put $j = k$ and perform a sum over the Cartesian subscripts. We obtain

$$\sigma = \delta + 3\varepsilon. \qquad (13.44)$$

Then we multiply Eq. (13.43) by $v_j v_k$ and sum again. We arrive at

$$\sigma^{(2)} = \delta + \varepsilon. \qquad (13.45)$$

Solving these two equations for $\delta$ and $\varepsilon$, we immediately obtain Eq. (13.41) from Eq. (13.43).

The distribution functions $\Delta_\omega^{(l)}(\mathbf{r}, \mathbf{v})$ with $l \geq 2$ may be derived in an analogous manner. It is, however, again advisable to proceed in the reverse direction if no terms with derivatives of higher than the fourth order are to be obtained. Eq. (13.23) is immediately generalized to yield

$$2|\omega| \oint \Delta_\omega^{(l)}(\mathbf{r}, \mathbf{v}) \, d\Omega = -\oint \mathbf{v} \cdot \tilde{\partial} \Delta_\omega^{(l-1)}(\mathbf{r}, \mathbf{v}) \, d\Omega. \qquad (13.46)$$

The contribution with $l = 3$ vanishes since the integral on the righthand side contains an odd number (three) of vectors $\mathbf{v}$. In order to obtain the contribution with $l = 4$, we further derive

$$(2|\omega| + nv\sigma_{tr}) \oint v \cdot \tilde{\partial} \Delta_\omega^{(3)}(r, v) \, d\Omega = -\oint (v \cdot \tilde{\partial})^2 \, \Delta_\omega^{(2)}(r, v) \, d\Omega, \qquad (13.47)$$

with $\sigma_{tr}$ from Eq. (13.22), from Eq. (13.16). Since it is

$$\oint v_j v_k v_l v_m \, d\Omega = \frac{4\pi}{15} \delta_{jklm} \qquad (13.48)$$

with

$$\delta_{jklm} := \delta_{jk}\delta_{lm} + \delta_{jl}\delta_{km} + \delta_{jm}\delta_{kl}, \qquad (13.49)$$

we finally derive

$$\oint \Delta_\omega^{(4)}(r, v) \, d\Omega = \frac{1}{2|\omega| \, (2|\omega| + nv\sigma_{tr})^2} \frac{4\pi v^4}{15}$$
$$\cdot (b\delta_{jklm} + 5c\delta_{jk}\delta_{lm}) \, \tilde{\partial}_j \tilde{\partial}_k \tilde{\partial}_l \tilde{\partial}_m \Delta_\omega^{(0)}(r). \qquad (13.50)$$

Here, use was also made of Eqs. (13.46), (13.47), (13.38), and (13.6). In this way we arrive at

$$\left[ \ln\left(\frac{T_c}{T}\right) + \frac{v^2}{6} \pi T \sum_\omega \frac{1}{|\omega|^2(2|\omega| + nv\sigma_{tr})} \tilde{\partial} \cdot \tilde{\partial} \right.$$

$$+ \frac{v^4}{30} \pi T \sum_\omega \frac{\delta_{jklm} + \frac{5nv(\sigma - \sigma^{(2)})}{4|\omega|} \delta_{jk}\delta_{lm}}{|\omega|^2(2|\omega| + nv\sigma_{tr})^2 \, (2|\omega| + \frac{3}{2}nv(\sigma - \sigma^{(2)}))} \qquad (13.51)$$

$$\left. \cdot \tilde{\partial}_j \tilde{\partial}_k \tilde{\partial}_l \tilde{\partial}_m + \cdots \right] \Delta(r) = 0$$

which essentially is a generalization, including higher order derivatives, of Eq. (13.25). Eq. (13.9) was used, the cut-off at the other sums neglected. For position dependent impurity concentration, we would instead have found

$$\left[ \ln\left(\frac{T_c}{T}\right) + \frac{v^2}{6} \pi T \sum_\omega \frac{1}{|\omega|^2} \tilde{\partial} \cdot \frac{1}{2|\omega| + nv\sigma_{tr}} \tilde{\partial} \right.$$

$$+ \frac{v^4}{30} \pi T \delta_{jklm} \sum_\omega \frac{1}{|\omega|^2} \tilde{\partial}_j \frac{1}{2|\omega| + nv\sigma_{tr}} \tilde{\partial}_k \frac{1}{2|\omega| + \frac{3}{2}nv(\sigma - \sigma^{(2)})} \qquad (13.52)$$

$$\cdot \tilde{\partial}_l \frac{1}{2|\omega| + nv\sigma_{tr}} \tilde{\partial}_m$$

$$+ \frac{v^4}{24} \pi T \sum_\omega \frac{1}{|\omega|^3} \left( \tilde{\partial} \cdot \frac{1}{2|\omega| + nv\sigma_{tr}} \tilde{\partial} \right) \frac{nv(\sigma - \sigma^{(2)})}{2|\omega| + \frac{3}{2}nv(\sigma - \sigma^{(2)})}$$

$$\left. \cdot \left( \tilde{\partial} \cdot \frac{1}{2|\omega| + nv\sigma_{tr}} \tilde{\partial} \right) + \cdots \right] \Delta(r) = 0.$$

The clean limit $(n = 0)$ is easily derived from Eq. (13.51). Expressing the $\omega$ sums by Riemann's zeta function, we obtain

$$\left[\ln\left(\frac{T_c}{T}\right) + \frac{7\zeta(3)}{12}\,\xi_0^2(T)\,\tilde{\partial}\cdot\tilde{\partial} + \frac{31\zeta(5)}{240}\,\xi_0^4(T)\,\delta_{jklm}\tilde{\partial}_j\tilde{\partial}_k\tilde{\partial}_l\tilde{\partial}_m + \cdots\right]\Delta(r) = 0 \tag{13.53}$$

with $\zeta(3)$ given in Eq. (13.12) and

$$\zeta(5) = 1.0369\ldots. \tag{13.54}$$

The length $\xi_0(T)$ was given in Eq. (13.11). It is convenient to make use of

$$\delta_{jklm}\tilde{\partial}_j\tilde{\partial}_k\tilde{\partial}_l\tilde{\partial}_m = 3[(\tilde{\partial}\cdot\tilde{\partial})^2 + (2eH(r))^2] \tag{13.55}$$

in Eqs. (13.51) and (13.53). This relation is easily derived with the help of Eq. (11.57). Eq. (13.51) has — apart from a simplification which is valid for isotropic scattering cross section — first been given by *Tewordt* [43].

Some care is required in applying the generalized GL equations (13.51) through (13.53). The approximation scheme is only valid for sufficiently slowly varying functions $\Delta(r)$. But the higher order differential equations derived above also admit other solutions but slowly varying ones. The situation is rather simple for a homogeneous magnetic field and homogenous impurity concentration (if any). Eq. (13.1) is generalized to

$$\tilde{\partial}\cdot\tilde{\partial}\Delta(r) = -\lambda(T, H)\,\Delta(r) \tag{13.56}$$

with

$$\ln\left(\frac{T_c}{T}\right) - \lambda\,\frac{v^2}{6}\,\pi T \sum_\omega \frac{1}{|\omega|^2\,(2|\omega| + nv\sigma_{\mathrm{tr}})}$$
$$+ \frac{v^4}{30}\,\pi T \sum_\omega \frac{3(\lambda^2 + (2eH)^2) + \dfrac{5nv(\sigma - \sigma^{(2)})}{4|\omega|}\,\lambda^2}{|\omega|^2(2|\omega| + nv\sigma_{\mathrm{tr}})^2(2|\omega| + \tfrac{3}{2}nv(\sigma - \sigma^{(2)}))} + \cdots = 0. \tag{13.57}$$

Only that solution $\lambda$ is admitted which is obtained from Eq. (13.57) in a recursive manner. We give this solution only for a clean metal,

$$\lambda(T, H)\,\xi_0^2(T) = \frac{12}{7\zeta(3)}\ln\left(\frac{T_c}{T}\right) + \frac{31\,\zeta(5)}{10}\left(\frac{6}{7\,\zeta(3)}\right)^3 \ln^2\left(\frac{T_c}{T}\right)$$
$$+ \frac{93\,\zeta(5)}{140\,\zeta(3)}\,(2eH\,\xi_0^2(T))^2 + \cdots$$
$$= \frac{12}{7\,\zeta(3)}\,\frac{T_c - T}{T_c} + \left[\frac{6}{7\,\zeta(3)} + \frac{31\,\zeta(5)}{10}\left(\frac{6}{7\,\zeta(3)}\right)^3\right]\left(\frac{T_c - T}{T_c}\right)^2 \tag{13.58}$$
$$+ \frac{93\,\zeta(5)}{140\,\zeta(3)}\,(2eH\,\xi_0^2(T))^2 + \cdots.$$

The last expression is obtained by expanding the logarithm. If magnetic field or impurity concentration or both are position dependent, a more sophisticated perturbation treatment is required.

## 14. Diffusion Approximation

An approximation to the BCS-Gorkov theory of superconductivity which is expected to be valid for sufficiently short electronic mean free path was introduced in 1964 by *Maki* [9] and *de Gennes* [10]. This approximation shall in the present section be derived [7] with the help of the method of the correlation function.

### a) Derivation and Discussion of the Diffusion Equation

For sufficiently short mean free path in a sense to be specified below [Eq. (14.15)], the distribution function [Eq. (11.22)] is expected to be almost isotropic with respect to the direction of $v$. Therefore, we make the ansatz

$$\Delta_\omega(r, v) = H_\omega(r) + v \cdot H_\omega(r) + \cdots \tag{14.1}$$

and neglect contributions corresponding to higher harmonics. Obviously, $H_\omega(r)$ is related to the density of electrons (or of carriers of information) in real space. Indeed it is

$$\oint \Delta_\omega(r, v) \, d\Omega = 4\pi \, H_\omega(r). \tag{14.2}$$

$H_\omega(r)$ is related to the particle flux since it is

$$\oint v \Delta_\omega(r, v) \, d\Omega = \frac{4\pi}{3} \, v^2 H_\omega(r); \tag{14.3}$$

cf. Eq. (13.6).

Now Eq. (14.1) is inserted into Eq. (11.23) and the resulting equation is integrated over all directions of $v$ with the help of Eq. (13.6). We obtain

$$2|\omega| \, H_\omega(r) + \frac{v^2}{3} \tilde{\partial} \cdot H_\omega(r) = \frac{1}{(2\pi)^2} \, \Delta(r). \tag{14.4}$$

This equation is nothing but the Laplace transformed continuity equation. It shall be discussed on a more general basis in Section 14d.

If Eq. (11.23), with Eq. (14.1) inserted into it, is first multiplied by $v$ and afterwards integrated over all directions of the velocity vector,

$$(2|\omega| + n v \sigma_{tr}) \, H_\omega(r) + \tilde{\partial} H_\omega(r) = 0 \tag{14.5}$$

is obtained. Now also Eqs. (13.19) and (13.22) were applied. Eq. (14.5) may be approximated by

$$n v \sigma_{tr} H_\omega(r) + \tilde{\partial} H_\omega(r) = 0 \tag{14.6}$$

for sufficiently big concentration of impurities (sufficiently short mean free path). But Eq. (14.6) is certainly not correct for arbitrarily big values of $|\omega|$. Actually not even the ansatz (14.1) is justified in this case. We mention that Eq. (14.6) may be interpreted as Fick's law in the theory of diffusion. Indeed,

$$D = \frac{v l_{tr}}{3} \tag{14.7}$$

is the diffusion coefficient of the kinetic theory. Here, the transport length was introduced from Eq. (13.27).

The function $H_\omega(r)$ may be eliminated from Eqs. (14.4) and (14.6). The result is

$$\left( 2 |\omega| - \frac{v l_{tr}}{3} \tilde{\partial} \cdot \tilde{\partial} \right) H_\omega(r) = \frac{1}{(2\pi)^2} \Delta(r). \tag{14.8}$$

This equation may, because of Eq. (14.7), be interpreted as the Laplace transformed diffusion equation. For the complete solution of the problem also the gap equation (10.1) has to be solved which may now be written in the form

$$\Delta(r) = (2\pi)^3 g(r) N(r) T \sum_\omega{}' H_\omega(r); \tag{14.9}$$

cf. Eqs. (13.5) and (14.2).

A diffusion equation for the kernel is derived from Eq. (14.8) with the help of Eqs. (11.24) and (14.2). We obtain

$$\left( 2 |\omega| - \frac{v l_{tr}}{3} \tilde{\partial} \cdot \tilde{\partial} \right) K_\omega(r, r') = 2\pi N \delta(r - r'), \tag{14.10}$$

where $N$ is the density of states.

If the transport length, $l_{tr}$, is position dependent, the replacement

$$2 |\omega| - \frac{v l_{tr}}{3} \tilde{\partial} \cdot \tilde{\partial} \to 2 |\omega| - \tilde{\partial} \cdot \frac{v l_{tr}}{3} \tilde{\partial} \tag{14.11}$$

has to be made both in Eq. (14.8) and Eq. (14.10). This follows immediately from the derivation given above. Finally, for paramagnetic impurities, we have to replace

$$2 |\omega| - \frac{v l_{tr}}{3} \tilde{\partial} \cdot \tilde{\partial} \to 2(|\omega| + n v \sigma_{sp}) - \frac{v l_{tr}}{3} \tilde{\partial} \cdot \tilde{\partial} \tag{14.12}$$

8*

with $\sigma_{sp}$ from Eq. (10.15) and $l_{tr}$ calculated with the help of Eqs. (13.27) and (13.37).

The differential equations have to be supplemented by boundary conditions on surfaces and interfaces which shall be derived in Section 16. In an infinite metal we postulate that $K_\omega(r, r')$ is bounded (and actually vanishes) at infinity. Eq. (14.10) is then easily solved in the absence of a magnetic field $[A(r) \equiv 0]$ with the result

$$K_\omega(r, r') = \frac{3N}{2v l_{tr} |r - r'|} \exp\left[-\sqrt{\frac{3}{l_{tr} \xi_\omega}} |r - r'|\right]. \qquad (14.13)$$

It is seen that the range of $K_\omega(r, r')$ is roughly given by $\sqrt{\xi_\omega l_{tr}}$. This result was used in Section 13b for the discussion of the validity of the GL approximation in the presence of impurities.

The degree of singularity at $r \approx r'$ in Eq. (14.13) is unexpected. It is seen from Eq. (12.17) that the particles starting from $r'$ always produce a singularity $\propto |r - r'|^{-2}$. Quite generally it is

$$\lim_{r \to r'} (r - r')^2 K_\omega(r, r') = \frac{N}{2v}. \qquad (14.14)$$

The reason that this stronger singularity does not show up in Eq. (14.13) is that the distribution function of those particles, which have not suffered a single impurity collision, is strongly anisotropic so that it is not covered by the almost isotropic ansatz (14.1). Nevertheless, the solution (14.13) satisfies the sum rule (3.13).

A condition for the validity of the diffusion approximation is in this way easily obtained: the pair potential has to vary slowly on $l_{tr}$ (or on $l$), the length on which the distribution function of not yet scattered particles is important. This condition shall be put on a more quantitative basis in Section 14c. The result will be that the diffusion approximation is valid in the so-called dirty limit,

$$l_{tr} \ll \xi_0(T_c), \qquad (14.15)$$

with $\xi_0(T)$ given in Eq. (13.11). The approximation is justified not only for $T$ close to $T_c$ but at all temperatures between 0 and $T_c$ though the condition (14.15) has definitely to be formulated with $\xi_0(T_c)$ and not with $\xi_0(T)$. It will turn out in Section 16 that there is a further requirement in finite samples: all linear dimensions have to be big as compared to $l_{tr}$. It should be kept in mind that a further problem of validity is connected with essentially the transition from Eq. (14.5) to Eq. (14.6). This problem is postponed to the end of Section 14c.

## b) General Properties of the Kernel

The uniqueness theorem and the general properties of the kernel (Sections 3, 5, and 11) may also be proved within the present approximate scheme.

For the proof of the uniqueness theorem, it is assumed that $K_\omega^{(1)}(r, r')$ and $K_\omega^{(2)}(r, r')$ are two solutions to Eq. (14.10), eventually with the substitution (14.11) made. It is to be shown that the difference,

$$F_\omega(r) = K_\omega^{(1)}(r, r') - K_\omega^{(2)}(r, r'), \tag{14.16}$$

vanishes for $|\omega| \neq 0$. For the proof, the homogeneous diffusion equation for $F_\omega(r)$ is multiplied by $F_\omega^*(r)$ and by the reciprocal density of states in the particular metal. The result is integrated over the whole sample. An integration by parts is performed in the derivative term. The result is

$$\sum_j N_j^{-1} \int_{V_j} \left[ 2|\omega| \, |F_\omega(r)|^2 + \frac{v l_{\mathrm{tr}}}{3} |\tilde\partial F_\omega(r)|^2 \right] d^3r \leq 0 \tag{14.17}$$

provided that it is

$$\sum_j N_j^{-1} \oint_{B_j} (v l_{\mathrm{tr}})_j \, F_{j\omega}^*(r) \, d^2r \cdot \tilde\partial F_{j\omega}(r) \leq 0 \tag{14.18}$$

as a consequence of the boundary conditions. $F_{j\omega}(r)$ is the boundary value of the function $F_\omega(r)$ in metal $j$. The other symbols were already explained in connection with Eq. (11.14). $F_\omega(r) \equiv 0$ follows immediately from Eq. (14.17). With the help of this uniqueness theorem it is easily shown that $K_\omega(r, r')$ is real for $A(r) \equiv 0$ provided that also the boundary conditions contain only real coefficients.

The Hermiticity (11.16) is proved by first multiplying Eq. (14.10) by $K_\omega^*(r, r'')$ and by the reciprocal of the density of states. The complex conjugated equation is subtracted with $r'$ and $r''$ interchanged. The difference is then integrated over the sample. Eq. (11.16) is obtained with a slightly different notation provided that it is

$$\sum_j N_j^{-1} \oint_{B_j} (v l_{\mathrm{tr}})_j \, d^2r \cdot \{K_{j\omega}^*(r, r'') \, \tilde\partial K_{j\omega}(r, r') - K_{j\omega}(r, r') \, (\tilde\partial K_{j\omega}(r, r''))^*\} = 0 \tag{14.19}$$

as a consequence of the boundary conditions.

The derivation of the lower and upper bounds of the operator $\mathbb{K}_\omega$ [Eqs. (11.19) and (11.33)] is based upon

$$\mathbb{K}_\omega \Delta(r) = (2\pi)^3 N(r) H_\omega(r), \tag{14.20}$$

which follows from Eqs. (11.24) and (14.2). Eq. (14.8) is now multiplied by $H_\omega^*(r)$ and by the density of states. The same manipulations as in connection with the uniqueness proof given above lead to

$$2|\omega| \sum_j N_j \int_{V_j} |H_\omega(r)|^2 \, d^3r \leq \frac{1}{(2\pi)^5} (\Delta, \mathbb{K}_\omega \Delta), \tag{14.21}$$

where another non-negative term on the lefthand side, containing gauge invariant derivatives, has been discarded. Eq. (14.21) is closely related to Eq. (11.35). It is indeed easily shown from Eq. (14.20) that it is

$$\sum_j N_j \int_{V_j} |H_\omega(r)|^2 \, d^3r = \frac{1}{(2\pi)^6} \left( \Delta, \mathbb{K}_\omega \frac{1}{\mathbb{N}} \mathbb{K}_\omega \Delta \right) \tag{14.22}$$

with the density of states operator given in Eq. (11.34). The further steps of the proof are the same ones as in Section 11b.

The sum rule (3.13) was already pointed out above in connection with Eq. (14.13) for an infinite sample consisting of a single metal. The general proof is obtained by integrating Eq. (14.10), for $A(r) \equiv 0$, with respect to $r$ over the whole sample. The result is

$$2|\omega| \int K_\omega(r, r') \, d^3 r = 2\pi N(r) , \tag{14.23}$$

provided that it is

$$\sum_j \oint_{B_j} (v l_{\text{tr}})_j \, d^2 r \cdot \tilde{\partial} K_{j\omega}(r, r') = 0 \tag{14.24}$$

as a consequence of the boundary conditions.

## c) Ginzburg-Landau-like Formulation

The Ginzburg-Landau (GL) equation (13.1) is a differential equation for the pair potential, $\Delta(r)$, whereas the diffusion equation (14.10) is a differential equation for the kernel, $K_\omega(r, r')$. It has, however, been shown by *de Gennes* [10] that the diffusion approximation may be cast into a GL-like formulation if the sample consists of a single metal.

In order to derive this formulation, we start from the complete orthonormal set of eigenfunctions, $\varphi_m(r)$, to the equation

$$\tilde{\partial} \cdot \tilde{\partial} \, \varphi_m(r) = -\lambda_m \varphi_m(r) . \tag{14.25}$$

In an infinite sample it is required that the solutions be bounded at infinity. They are actually not normalizable in this case but this complication shall be disregarded. In a finite sample, the boundary condition (16.4) has to be satisfied on the surface.

It follows from the postulated completeness and orthonormality that it is

$$\sum_m \varphi_m(r) \, \varphi_m^*(r') = \delta(r - r') . \tag{14.26}$$

Expanding $K_\omega(r, r')$ into the complete set $\{\varphi_m(r)\}$ with coefficients depending upon $r'$ and inserting this expansion into Eq. (14.10), we find

$$K_\omega(r, r') = 2\pi N \sum_m \frac{\varphi_m(r) \, \varphi_m^*(r')}{2|\omega| + \dfrac{v l_{\text{tr}}}{3} \lambda_m} . \tag{14.27}$$

The functions $\varphi_m(r)$ are simultaneous eigenfunctions to all kernels $K_\omega(r, r')$,

$$\int K_\omega(r, r') \, \varphi_m(r') \, d^3 r' = \frac{2\pi N}{2|\omega| + \dfrac{v l_{\text{tr}}}{3} \lambda_m} \, \varphi_m(r) . \tag{14.28}$$

They also solve the gap equation (10.1) provided that it is

$$1 = 2\pi g N T \sum_\omega{}' \frac{1}{2|\omega| + \dfrac{v l_{\text{tr}}}{3} \lambda_m} . \tag{14.29}$$

It will turn out below that the transition temperature following from Eq. (14.29) is biggest for the smallest possible $\lambda_m$. Therefore, the pair potential is put equal (or proportional) to a $\varphi_m(r)$ with smallest eigenvalue. It satisfies the linearized GL equation (13.1) with $\lambda(T)$ being implicitly defined by

$$1 = 2\pi g N T \sum_{\omega}' \frac{1}{2|\omega| + \dfrac{v l_{\mathrm{tr}}}{3}\lambda}. \tag{14.30}$$

Whereas the GL equation proper is valid only for temperatures close to $T_c$ the present GL version of the diffusion approximation is valid in the dirty limit [Eq. (14.15)] at all temperatures.

It is not necessary that the impurity concentration (or the transport length) is position independent. Otherwise, also $\lambda$, as defined by Eq. (14.30), will become position dependent. It is obviously proportional to $l_{\mathrm{tr}}^{-1}$. Eq. (13.35) is the appropriate GL equation in this case.

Eq. (14.30) may be simplified with the help of Eq. (13.9). In this way we obtain

$$\ln\left(\frac{T_c}{T}\right) = 2\pi T \sum_{\omega}\left(\frac{1}{2|\omega|} - \frac{1}{2|\omega| + \dfrac{v l_{\mathrm{tr}}}{3}\lambda}\right). \tag{14.31}$$

The cut-off shall be disregarded which is allowed for

$$\frac{v l_{\mathrm{tr}}\lambda}{3} \ll 2\omega_{\mathrm{D}}. \tag{14.32}$$

It will turn out below that Eq. (14.32) is always satisfied in the weak-coupling case (11.51).

In order to simplify the notation we now introduce a function, $\mu(T)$, implicitly by

$$\ln\left(\frac{T_c}{T}\right) = 2\pi T \sum_{\omega}\left(\frac{1}{2|\omega|} - \frac{1}{2|\omega| + \mu}\right). \tag{14.33}$$

Comparison with Eq. (14.31) shows that it is

$$\mu(T) = \frac{v l_{\mathrm{tr}}}{3}\lambda(T). \tag{14.34}$$

The paramagnetic impurity problem of Section 11c is contained in the present formulation if we put

$$\mu = 2 n v \sigma_{\mathrm{sp}}; \tag{14.35}$$

cf. Eq. (11.60). Eq. (14.33) may be rewritten in the form

$$\ln\left(\frac{T_c}{T}\right) = 2 \sum_{n=0}^{\infty} \left( \frac{1}{2n+1} - \frac{1}{2n+1+\dfrac{\mu}{2\pi T}} \right) \qquad (14.36)$$

or, with the help of the digamma function [Eq. (13.33)], in the form

$$\ln\left(\frac{T_c}{T}\right) = \psi\left(\frac{1}{2} + \frac{\mu}{4\pi T}\right) - \psi\left(\frac{1}{2}\right). \qquad (14.37)$$

$\mu(T)$ is a unique function of $T$ in the interval $0 \leq T \leq T_c$. It increases monotonically with decreasing temperature. It further is

$$\mu(T_c) = 0, \qquad (14.38)$$

$$\mu(T) = \frac{8}{\pi}(T_c - T) \qquad (14.39)$$

for $T \approx T_c$, and

$$\mu(0) = \frac{\pi T_c}{\gamma} \qquad (14.40)$$

with $\gamma$ given in Eq. (11.43). Because of

$$\mu(T) \leq \mu(0) \quad \text{for} \quad 0 \leq T \leq T_c, \qquad (14.41)$$

the neglection of the cut-off in Eqs. (14.31), (14.33), and (14.36) is certainly justified in the weak-coupling case (11.51). The proofs for the various statements will be supplied below in small print. $\mu/T_c$ is shown in Fig. 2 as a function of $T/T_c$.

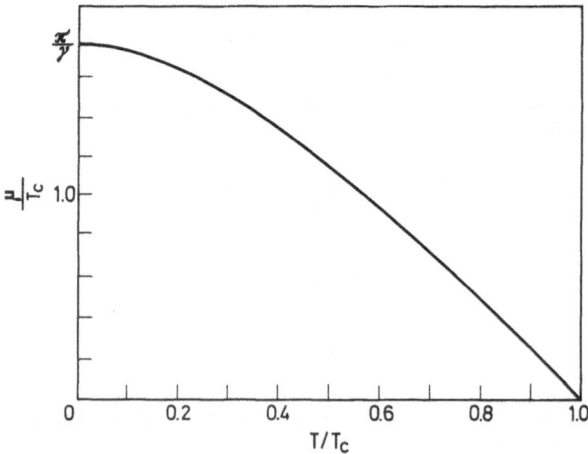

Fig. 2. The function $\mu(T)$ as defined implicitly by Eq. (14.33). The ordinate shows $\mu/T_c$, the abscissa $T/T_c$

Calculating $\lambda(T)$ for $T$ close to $T_c$ with the help of Eqs. (14.34) and (14.39), we obtain the GL result (13.26) in the dirty limit (13.29). Therefore, GL approximation and GL version of the diffusion approximation become identical in the region of common validity of both approximations.

It further follows from Eqs. (14.34), (14.41), and (14.40) that it is

$$\lambda(T) \leqq \frac{3}{2\gamma} \frac{1}{\xi_0(T_c) \, l_{\text{tr}}} \tag{14.42}$$

at an arbitrary temperature below $T_c$. With $\xi(T)$ defined by Eq. (13.2), it therefore is

$$\xi(T) \geqq \sqrt{\frac{2\gamma}{3} \xi_0(T_c) \, l_{\text{tr}}} \; . \tag{14.43}$$

The length $\xi(T)$ has to be big as compared to $l_{\text{tr}}$ in order that the diffusion approximation be valid; cf. the end of Section 14a. It is seen that the diffusion approximation is valid under this aspect in the dirty limit (14.15). It seems that even

$$\sqrt{\frac{l_{\text{tr}}}{\xi_0(T_c)}} \ll 1 \tag{14.44}$$

has to be postulated. Though the present considerations can be made only if the sample consists of a single metal, the validity requirement (dirty limit) is assumed to be the same for samples consisting of several metals in electric contact.

Applying the present results to the paramagnetic impurity case of Section 11c – where no dirty limit requirements have to be made –, we see from Eqs. (14.35) and (14.40) that the so-called critical concentration is given by [36]

$$n_c = \frac{\pi T_c}{2\gamma v \sigma_{\text{sp}}} \; . \tag{14.45}$$

The transition temperature becomes equal to zero for $n = n_c$.

We show that Eq. (14.36) defines $\mu \geqq 0$ uniquely. Indeed, if there were two solutions, $\mu_1$ and $\mu_2$, with

$$\mu_1 > \mu_2 \geqq 0 \tag{14.46}$$

for given $T > 0$, it would follow

$$\sum_{n=0}^{\infty} \left( \frac{1}{2n+1+\dfrac{\mu_2}{2\pi T}} - \frac{1}{2n+1+\dfrac{\mu_1}{2\pi T}} \right) = 0 \; . \tag{14.47}$$

This, however, is impossible since every term in the sum is positive.

Further we show that it is

$$\frac{d\mu(T)}{dT} < 0 \tag{14.48}$$

for $\mu \geq 0$, $T > 0$. It follows from Eq. (14.36) that it is

$$\frac{d\mu}{dT} = \frac{\mu}{T} - \frac{1}{F\left(\dfrac{\mu}{T}\right)} \tag{14.49}$$

with

$$F\left(\frac{\mu}{T}\right) = \frac{1}{\pi} \sum_{n=0}^{\infty} \frac{1}{\left(2n+1+\dfrac{\mu}{2\pi T}\right)^2} \, . \tag{14.50}$$

Eq. (14.48) is immediately derived from Eq. (14.49) since it is

$$0 < F\left(\frac{\mu}{T}\right) < \frac{T}{\mu} \, . \tag{14.51}$$

The first inequality sign is obvious. The second one is correct since from

$$\frac{1}{\left(2n+1+\dfrac{\mu}{2\pi T}\right)^2} < \int_{n-\frac{1}{2}}^{n+\frac{1}{2}} \frac{dn'}{\left(2n'+1+\dfrac{\mu}{2\pi T}\right)^2} \tag{14.52}$$

it follows that it is

$$F\left(\frac{\mu}{T}\right) < \int_{-\frac{1}{2}}^{\infty} \frac{dn}{\left(2n+1+\dfrac{\mu}{2\pi T}\right)^2} = \frac{T}{\mu} \, . \tag{14.53}$$

Eq. (14.38) is an obvious consequence of Eq. (14.36). Eq. (14.48) then leads to the statement that it is $\mu(T) > 0$ for $0 \leq T < T_c$. Eq. (14.39) is obtained from Eq. (14.36) by expanding both the logarithm for $T \approx T_c$ and the second term on the righthand side for $\mu/2\pi T \ll 1$. Eq. (13.30) is applied. For $T \to 0$ we may rewrite Eq. (14.36) in the form

$$\lim_{T \to +0}\left[\ln\left(\frac{T_c}{T}\right) - \lim_{c \to +\infty} 2 \sum_{n=0}^{c/T}\left(\frac{1}{2n+1} - \frac{1}{2n+1+\dfrac{\mu}{2\pi T}}\right)\right] = 0 \, . \tag{14.54}$$

The first sum is evaluated with the help of Eq. (11.42). The second sum is transformed into an integral,

$$2 \sum_{n=0}^{c/T} \frac{1}{2n+1+\dfrac{\mu}{2\pi T}} = 2 \int_0^{2\pi c} \frac{d\omega}{\omega + \dfrac{\mu}{2}} + \frac{\pi T}{4\pi c + \mu} + \cdots \, . \tag{14.55}$$

Eq. (14.40) is then easily derived.

There is still a problem connected with contributions from big values of $|\omega|$ for which $2|\omega| \ll n v \sigma_{tr}$ is not valid [cf. Eqs. (14.5) and (14.6)]. It will be shown in Section 17 [Eq. (17.27)] that Eq. (14.31) has to be replaced by

$$\ln\left(\frac{T_c}{T}\right) = 2\pi T \sum_{\omega}\left(\frac{1}{2|\omega|} - \frac{1}{2|\omega| + \dfrac{\lambda \zeta_\omega v}{3}}\right) \tag{14.56}$$

for arbitrary values of $|\omega|$, homogeneous impurity concentration, and isotropic scattering cross section. In this case, the transport length, $l_{tr}$, and the electronic mean free path, $l$, become equal. No statement of the kind of Eq. (14.56) can be made in other cases. The terms on the righthand side of Eqs. (14.31) and (14.56) are almost equal as long as $\zeta_\omega$ [Eq. (12.14)] is practically equal to $l$. This is certainly true for sufficiently small values of $|\omega|$ as a consequence of Eq. (14.15). Substantial deviations occur for $2|\omega| \gtrsim nv\sigma$. But then $\lambda l v/3$ and $\lambda \zeta_\omega v/3$ are both already small as compared to $2|\omega|$.

## d) Elimination of the Cut-off

At various occasions in Sections 13 and 14, the cut-off at $\pm \omega_D$ was eliminated with the help of Eq. (13.9). It shall now be shown that this elimination may be made directly [5] in the gap equation (10.1).

We introduce the flux vector

$$I_\omega(r, r') = 2\pi^2 N \oint v\, g_\omega(r, v; r')\, d\Omega . \tag{14.57}$$

If $K_\omega(r, r')$, with respect to the variable $r$, represents the density of electrons or of carriers of information, $I_\omega(r, r')$ is the corresponding particle flux. The relation

$$2|\omega|\, K_\omega(r, r') + \tilde{\partial} \cdot I_\omega(r, r') = 2\pi N(r)\, \delta(r - r') \tag{14.58}$$

is easily derived from the Boltzmann equation (10.5). In the presence of paramagnetic impurities, we find

$$2(|\omega| + nv\sigma_{sp})\, K_\omega(r, r') + \tilde{\partial} \cdot I_\omega(r, r') = 2\pi N(r)\, \delta(r - r') \tag{14.59}$$

instead. Eqs. (14.58) and (14.4) are essentially identical relations. For $A(r) \equiv 0$, Eq. (14.58) may be interpreted as the Laplace transformed continuity equation for the particle density (with the initial condition as inhomogeneity on the righthand side). In this case, the sum rule (3.13) is immediately derived since it is

$$\sum_j \oint_{B_j} d^2r \cdot I_\omega(r, r') = 0 \tag{14.60}$$

as a consequence of the boundary conditions. The contributions from external surfaces to the righthand side of Eq. (14.60) vanish because of Eq. (10.21). The contributions from interfaces vanish because of the analogous statement that the normal component of the particle flux is continuous on interfaces.

Going over to the complex conjugate of Eq. (14.58), interchanging $r$ and $r'$ and making use of Eq. (11.16) we derive

$$2|\omega|\, K_\omega(r, r') + [\tilde{\partial}' \cdot I_\omega(r', r)]^* = 2\pi N(r)\, \delta(r - r') . \tag{14.61}$$

Here, $\tilde{\partial}'$ denotes the gauge invariant derivative with respect to $r'$. The elimination of the cut-off is based upon this relation. Evidently it is

$$\int K_\omega(r, r')\, \Delta(r')\, d^3r' = \frac{\pi N(r)}{|\omega|}\, \Delta(r) + \frac{1}{2|\omega|} \int I_\omega^*(r', r) \cdot \tilde{\partial}' \Delta(r')\, d^3r' . \tag{14.62}$$

An integration by parts was here made. The boundary integral, which is produced by this integration, vanishes as a consequence of Eq. (14.60). Inserting Eq. (14.62) into Eq. (10.1) and making use of Eq. (13.9) we obtain

$$\ln\left(\frac{T_c}{T}\right) \Delta(r) = -\frac{1}{N}\, T \sum_\omega{}' \frac{1}{2|\omega|} \int I_\omega^*(r', r) \cdot \tilde{\partial}' \Delta(r')\, d^3r' . \tag{14.63}$$

If the cut-off in the sum may be disregarded – but this ought to be checked in every specific application –, the cut-off at the Debye frequency is in this way eliminated. Eq. (14.63) shall now be applied to the GL and diffusion approximations.

The distribution function $\Delta_\omega(r, v)$ was in the GL approximation derived in Section 13 b. Remembering the definition (11.22) of $\Delta_\omega(r, v)$, the results of Section 13 b may also be formulated as an expression for $g_\omega(r, v; r')$,

$$g_\omega(r, v; r') = \frac{1}{(2\pi)^2 \, 2|\omega|} \left(1 - \frac{1}{2|\omega| + nv\sigma_{tr}} v \cdot \tilde{\partial} + \cdots\right) \delta(r - r'). \qquad (14.64)$$

The flux vector is then calculated with the help of Eq. (14.57),

$$I_\omega(r', r) = -2\pi N \frac{v^2}{3(2|\omega|)(2|\omega| + nv\sigma_{tr})} \tilde{\partial}' \delta(r - r'). \qquad (14.65)$$

Taking into account

$$\tilde{\partial}' \delta(r - r') = -[\tilde{\partial}\delta(r - r')]^* \qquad (14.66)$$

we obtain

$$I_\omega^*(r', r) = +2\pi N \frac{v^2}{3(2|\omega|)(2|\omega| + nv\sigma_{tr})} \tilde{\partial}\delta(r - r'). \qquad (14.67)$$

Inserting Eq. (14.67) into Eq. (14.63) we deduce

$$\ln\left(\frac{T_c}{T}\right) \Delta(r) = -2\pi T \sum_\omega \frac{v^2}{3(2|\omega|)^2 (2|\omega| + nv\sigma_{tr})} \tilde{\partial} \cdot \tilde{\partial}\Delta(r), \qquad (14.68)$$

which is identical with the GL equation derived in Section 13 b. There it was shown that the cut-off may indeed be disregarded.

In the diffusion approximation, it is

$$I_\omega(r, r') = -\frac{vl_{tr}}{3} \tilde{\partial} K_\omega(r, r'). \qquad (14.69)$$

Whereas Eq. (14.58) corresponds to Eq. (14.4), Eq. (14.69) corresponds to Eq. (14.6); cf. also Eq. (14.7). The possible $r$ dependence of the transport length shall be disregarded though the final result (14.72) can be derived also in this more general case. It follows from Eq. (14.69) that it is

$$I_\omega^*(r', r) = -\frac{vl_{tr}}{3} [\tilde{\partial}' K_\omega(r', r)]^*. \qquad (14.70)$$

If the sample consists of a single metal – but not otherwise – an integration by parts may now be made in Eq. (14.63) without leading to a surface term. In this way, we obtain

$$\ln\left(\frac{T_c}{T}\right) \Delta(r) = -\frac{vl_{tr}}{3} \frac{1}{N} T \sum_\omega{}' \frac{1}{2|\omega|} \int K_\omega(r, r') \tilde{\partial}' \cdot \tilde{\partial}' \Delta(r') \, d^3 r'. \qquad (14.71)$$

The surface term vanishes in a finite sample because of Eq. (15.1). Also Eq. (11.16) was applied in deriving Eq. (14.71). Making now use of Eq. (13.1) and of Eq. (14.28) with $\lambda_m$ replaced by $\lambda$ and $\varphi_m(r)$ replaced by $\Delta(r)$, we arrive at

$$\ln\left(\frac{T_c}{T}\right) = 2\pi T \sum_\omega \frac{\dfrac{vl_{tr}}{3}\lambda}{2|\omega|\left(2|\omega| + \dfrac{vl_{tr}}{3}\lambda\right)}. \qquad (14.72)$$

This relation is identical with Eq. (14.31). It was shown in Section 14 c that the cut-off may again be disregarded.

### 15. Ginzburg-Landau Approximation: Boundary Conditions

In any finite sample and in any sample consisting of several metals, the linearized Ginzburg-Landau (GL) equation (Section 13) has to be supplemented by boundary conditions on surfaces and interfaces. Since the GL equation is a differential equation for the pair potential, $\Delta(r)$, the boundary conditions must be formulated in terms of $\Delta(r)$. *Ginzburg* and *Landau* themselves suggested [8]

$$n \cdot \tilde{\partial} \, \Delta(r) = 0 \qquad (15.1)$$

as the boundary condition on external surfaces. Here, $n$ is the normal unit vector on the surface in the point $r$.

In this section, boundary conditions on surfaces and interfaces shall be derived with the help of the method of correlation function. The boundary condition (15.1) will be confirmed in Section 15a for arbitrary metals (with spherical Fermi surface) and arbitrary reflexion properties of the surface. A microscopic derivation of Eq. (15.1) for clean metals and specular reflexion was published earlier by *Abrikosov* [31]. According to a remark in the literature [16], *Gorkov* gave a microscopic proof of the same boundary condition which has not been published. Boundary conditions on interfaces will also be derived (Section 15c) but this could so far only be done for clean metals. Also more refined boundary conditions on surfaces will be obtained (Section 15b) which should be used in connection with generalized GL equations. But also this task could so far be performed for clean metals only. Observational consequences of refined boundary conditions can, however, under certain conditions (single metal, homogeneous impurity concentration, isotropic scattering) be calculated for metals containing impurities with the help of a method to be developed in Section 17.

Eq. (15.1) is easily derived for specular reflexion. The GL approximation may be expressed in terms of a distribution function, $\Delta_\omega(r, v)$, which is a series expansion with respect to gauge invariant derivatives of the pair potential. According to Eqs. (13.15), (13.17), (13.21), and (13.22), it is

$$\Delta_\omega(r, v) = \frac{1}{(2\pi)^2 \, 2|\omega|} \left( 1 - \frac{1}{2|\omega| + nv\sigma_{\mathrm{tr}}} v \cdot \tilde{\partial} + \cdots \right) \Delta(r), \qquad (15.2)$$

where higher order derivatives were neglected. The semi-microscopic boundary condition of specular reflexion [Eqs. (10.16) and (10.22)] may now be written as

$$\Delta_\omega(r, v) = \Delta_\omega(r, v - 2n(n \cdot v)). \qquad (15.3)$$

The first term on the righthand side of Eq. (15.2) satisfies this boundary condition [and actually any boundary condition because of Eq. (10.20)]. The second term satisfies Eq. (15.3) if and only if Eq. (15.1) holds.

With the boundary condition (15.1), the GL equation (13.1) or (13.35) is solved by constant $\Delta(r)$ for $A(r) \equiv 0$. $T_c$ is the transition temperature also for finite samples in agreement with the general analysis of Section 11c. The transition temperature is always smaller than $T_c$ in the presence of a magnetic field. Indeed it follows from Eqs. (13.35), (13.26) and (15.1) that it is

$$\frac{T_c - T}{T_c} \int |\Delta(r)|^2 \, d^3r = \frac{1}{7\zeta(3)\,\xi_0^2(T_c)} \int \frac{1}{\chi\left(\dfrac{\xi_0(T_c)}{l_{tr}}\right)} |\tilde{\partial}\Delta(r)|^2 \, d^3r . \tag{15.4}$$

Since also the function $\chi(\xi_0/l_{tr})$ is non-negative, $T$ is smaller than $T_c$ if not Eq. (11.56) is satisfied. The further arguing proceeds along the same lines as in connection with Eq. (11.56).

## a) Surfaces, Lowest Order

A more refined way of arguing is required in all cases but the simple one of specular reflexion treated above. The GL distribution function (15.2) does not satisfy arbitrary boundary conditions (10.16). This is actually not surprising since the derivation of the GL equation given in Section 13 is not valid in the neighbourhood of surfaces and interfaces. It is, on the contrary, required that no surfaces and interfaces lie within the range of the kernel, $\sum' K_\omega(r, r')$. The GL approximation is, therefore, not valid in surface and interface layers of a thickness of this order of magnitude. The GL approximation is not valid at all if not all linear dimensions of the sample (including radii of curvature of surfaces and interfaces) are big as compared to the range of $\sum' K_\omega(r, r')$. This range can be given in a closed form only in limiting cases. It is of the order of $\xi_0(T)$ [Eq. (13.11)] in the clean limit and of the order of $\sqrt{\xi_0(T)\, l_{tr}}$ in the dirty limit. $T$ may be put equal to $T_c$ in these estimates since the GL approximation is only valid for temperatures in the neighbourhood of the transition temperature.

The problem of deriving boundary conditions to be used in conjunction with the GL equation can, therefore, only mean that a solution to the GL equation is extrapolated, through the surface and interface layers mentioned above, to the surfaces and interfaces. It is hoped that the "best" extrapolated solution will satisfy certain boundary conditions on surfaces and interfaces. The concept of "best" extrapolated solution gets its exact meaning in connection with the variational principle of Section 2b which is defined by the functional (2.20). Both the basic ideas and the detailed analysis were given by *Schöler* [12] in his unpublished thesis.

The functional (2.20) is required to be stationary with respect to arbitrary variation of the pair potential, $\Delta(r)$. It may be rewritten with the help of Eq. (11.24) to yield

$$\Phi = \int \frac{|\Delta(r)|^2}{g(r)} \, d^3r - \int d^3r \, 2\pi^2 N(r) \, \Delta^*(r) \, T \sum_\omega{}' \oint d\Omega \, \Delta_\omega(r, v) \, . \qquad (15.5)$$

Now we put

$$\Delta_\omega(r, v) = \Delta_{\omega 0}(r, v) + \Delta_{\omega 1}(r, v) \qquad (15.6)$$

with $\Delta_{\omega 0}(r, v)$ given by the GL approximation, i. e., by the righthand side of Eq. (15.2) including higher derivatives up to the desired order. The correction term, $\Delta_{\omega 1}(r, v)$, obeys a homogeneous Boltzmann equation,

$$(2|\omega| + v \cdot \tilde{\partial} + nv\sigma) \, \Delta_{\omega 1}(r, v) - nv \oint \frac{d\sigma(\theta)}{d\Omega} \, \Delta_{\omega 1}(r, v') \, d\Omega' = 0 \, , \qquad (15.7)$$

but an inhomogeneous semi-microscopic boundary condition of the form

$$\Delta_{\omega 1}(r, v_{\text{out}}) - \int R(r; v_{\text{out}}, v_{\text{in}}) \, \Delta_{\omega 1}(r, v_{\text{in}}) \, d\Omega_{\text{in}} = -\Delta_{\omega 0}(r, v_{\text{out}})$$
$$+ \int R(r; v_{\text{out}}, v_{\text{in}}) \, \Delta_{\omega 0}(r, v_{\text{in}}) \, d\Omega_{\text{in}} \qquad (15.8)$$

on external surfaces in order that the lefthand side of Eq. (15.6) satisfies the correct boundary condition. The biggest, velocity independent, term in $\Delta_{\omega 0}(r, v)$ satisfies the boundary condition because of Eq. (10.20). Therefore, $\Delta_{\omega 1}(r, v)$ may be regarded as a small quantity in the sense of the GL approximation. Further, $\Delta_{\omega 1}(r, v)$ is markedly different from zero only in the surface and interface layers mentioned above. It will be seen that its contribution to the righthand side of Eq. (15.5) may, to any desired accuracy, be transformed into a surface or interface integral.

At present, we want to derive boundary conditions to the ordinary linearized GL equation. In $\Delta_{\omega 0}(r, v)$ we, therefore, do not go beyond the second derivative so that it is

$$\oint d\Omega \Delta_{\omega 0}(r, v) = \frac{1}{\pi} \left[ \frac{1}{2|\omega|} + \frac{v^2}{3(2|\omega|)^2} \, \tilde{\partial} \cdot \frac{1}{2|\omega| + nv\sigma_{\text{tr}}} \, \tilde{\partial} + \cdots \right] \Delta(r) \, ; \qquad (15.9)$$

cf. Eqs. (13.15), (15.2), and (13.24). From Eq. (15.7), we obtain

$$2|\omega| \oint d\Omega \Delta_{\omega 1}(r, v) = -\oint d\Omega \, v \cdot \tilde{\partial} \Delta_{\omega 1}(r, v) \, . \qquad (15.10)$$

If this expression is inserted into Eqs. (15.5), an integration by parts may be performed with the result

$$
\int d^3 r \Delta^*(r) \oint d\Omega \Delta_{\omega 1}(r, v) = -\oint d\Omega \oint d^2 r \cdot v \Delta^*(r) \frac{1}{2|\omega|} \Delta_{\omega 1}(r, v)
$$

$$
+ \int d^3 r \oint d\Omega [v \cdot \tilde{\partial} \Delta(r)]^* \frac{1}{2|\omega|} \Delta_{\omega 1}(r, v) .
$$

(15.11)

The new volume integral may consistently be neglected since the integrand contains a factor $[v \cdot \tilde{\partial} \Delta(r)]^*$ and a factor $\Delta_{\omega 1}(r, v)$ which are both small in the sense of the GL approximation, Further the integrand is different from zero only in a thin surface layer. $\Delta_{\omega 1}(r, v)$ may, in the surface integral, be expressed by $\Delta_{\omega 0}(r, v)$ since it is

$$
\oint d\Omega \, n \cdot v [\Delta_{\omega 0}(r, v) + \Delta_{\omega 1}(r, v)] = 0
$$

(15.12)

as a consequence of the boundary condition; cf. Eq. (10.21). Inserting Eq. (15.2), we see that only the first derivative of $\Delta(r)$ contributes to the righthand side of Eq. (15.11).

Collecting now everything we obtain

$$
\Phi = \int \Delta^*(r) \left[ \frac{1}{g} - 2\pi N T \sum_{\omega}' \left( \frac{1}{2|\omega|} + \frac{v^2}{3(2|\omega|)^2} \, \tilde{\partial} \cdot \frac{1}{2|\omega| + n v \sigma_{tr}} \tilde{\partial} \right) \right] \Delta(r) \, d^3 r
$$

$$
+ 2\pi N \oint \Delta^*(r) \, T \sum_{\omega}' \frac{v^2}{3(2|\omega|)^2 (2|\omega| + n v \sigma_{tr})} d^2 r \cdot \tilde{\partial} \Delta(r) .
$$

(15.13)

An integration by parts leads to

$$
\Phi = \int \left[ \frac{|\Delta(r)|^2}{g} - 2\pi N T \sum_{\omega}' \left( \frac{1}{2|\omega|} |\Delta(r)|^2 \right. \right.
$$

$$
\left. \left. - \frac{v^2}{3(2|\omega|)^2 (2|\omega| + n v \sigma_{tr})} |\tilde{\partial} \Delta(r)|^2 \right) \right] d^3 r ,
$$

(15.14)

which shows that also the approximate $\Phi$ is a real functional of $\Delta(r)$. Therefore, $\Delta(r)$ and $\Delta^*(r)$ may be varied independently. Varying $\Delta^*(r)$ in Eq. (15.13), we obtain the GL equation (13.25) or its generalization to position dependent impurity concentration, from the volume integral and obtain the boundary condition (15.1) from the surface integral. It is remarkable that only general properties of the reflexion kernel were used in the derivation.

The extrapolation procedure referred to above was in the present treatment made by transforming the boundary layer term containing $\Delta_{\omega 1}(r, v)$ into a surface term proper.

## b) Surfaces, Higher Orders (for Clean Metals only)

For the more general analysis to be made now we rewrite Eqs. (15.5) and (15.6) in the form

$$\Phi = \Phi_0 + \Phi_1 \tag{15.15}$$

with

$$\Phi_0 = \int \frac{|\Delta(r)|^2}{g} \, d^3r - \int d^3r \, 2\pi^2 N(r) \, \Delta^*(r) \, T \sum_\omega{}' \oint d\Omega \Delta_{\omega 0}(r, v) \tag{15.16}$$

and

$$\begin{aligned}
\Phi_1 &= -\int d^3r \, 2\pi^2 N(r) \, \Delta^*(r) T \sum_\omega{}' \oint d\Omega \Delta_{\omega 1}(r, v) \\
&= 8\pi^4 N \oint d\Omega \oint d^2r \cdot v \, T \sum_\omega{}' \Delta^*_{\omega 0}(r, -v) \, \Delta_{\omega 1}(r, v).
\end{aligned} \tag{15.17}$$

If $\Delta_{\omega 0}(r, -v)$ is approximated by the $v$ independent term, the surface integral of Section 15a is again obtained.

The second equality sign in Eq. (15.17) is derived in the following way: Eq. (15.7) is multiplied by $\Delta^*_{\omega 0}(r, -v)$. Further, the complex conjugate of

$$(2|\omega| - v \cdot \tilde{\partial} + nv\sigma) \Delta_{\omega 0}(r, -v) - nv \oint \frac{d\sigma(\theta)}{d\Omega} \Delta_{\omega 0}(r, -v') \, d\Omega' = \frac{1}{(2\pi)^2} \Delta(r) \tag{15.18}$$

is multiplied by $\Delta_{\omega 1}(r, v)$. The difference of both equations obtained in this way is integrated with respect to $v$ over all directions and with respect to $r$ over the sample which is assumed to consist of a single metal. After an integration by parts only the surface integral in Eq. (15.17) survives.

In a clean metal ($n = 0$), Eq. (15.7) may easily be integrated along the straight classical orbits; cf. Section 12a. The modulus of $\Delta_{\omega 1}(r, v)$ decreases exponentially along the orbit with the decay length $\xi_\omega$. If all linear dimensions of the sample are big as compared to $\xi_0(T_c)$, we, therefore, may put

$$\Delta_{\omega 1}(r, v_{in}) = 0 \tag{15.19}$$

in every surface point $r$: practically no particles, having started from the opposite end of the straight orbit, reach the surface with velocity vector $v_{in}$. Eq. (15.19) is not true in the presence of impurities since an arbitrarily short orbit may be bent back to the boundary through a scattering event. As a consequence of Eq. (15.19), the boundary condition (15.8) is very much simplified,

$$\Delta_{\omega 1}(r, v_{out}) = -\Delta_{\omega 0}(r, v_{out}) + \int R(r; v_{out}, v_{in}) \Delta_{\omega 0}(r, v_{in}) \, d\Omega_{in}. \tag{15.20}$$

Eqs. (15.19) and (15.20) permit to express $\Delta_{\omega 1}(r, v)$ in Eq. (15.17) by $\Delta_{\omega 0}(r, v)$. Indeed, we obtain

$$\begin{aligned}
\Phi_1 = 8\pi^4 N \int d\Omega_{out} \int d\Omega_{in} \oint d^2r \cdot v_{out} R(r; v_{out}, v_{in}) \\
T \sum_\omega{}' \Delta^*_{\omega 0}(r, -v_{out}) \left[\Delta_{\omega 0}(r, v_{in}) - \Delta_{\omega 0}(r, v_{out})\right],
\end{aligned} \tag{15.21}$$

where use was also made of Eq. (10.20).

Eq. (13.15), with $\varDelta_\omega(r, v)$ replaced by $\varDelta_{\omega 0}(r, v)$, may also be used for clean metals. But now it simply is

$$\varDelta_{\omega 0}^{(l)}(r, v) = \frac{1}{(2\pi)^2 (2|\omega|)^{l+1}} (-v \cdot \tilde\partial)^l \varDelta(r); \qquad (15.22)$$

cf. Eq. (13.4). Inserting Eq. (13.15) into Eq. (15.21), we obtain

$$\varPhi_1 = \sum_{l, m=0}^{\infty} \varPhi_1(l, m) \qquad (15.23)$$

with $\varPhi_1(l, m)$ given by

$$\varPhi_1(l, m) = 8\pi^4 N \int d\Omega_{\text{out}} \int d\Omega_{\text{in}} \oint d^2 r \cdot v_{\text{out}} R(r; v_{\text{out}}, v_{\text{in}})$$
$$T \sum_\omega{}' \varDelta_{\omega 0}^{(l)*}(r, -v_{\text{out}}) [\varDelta_{\omega 0}^{(m)}(r, v_{\text{in}}) - \varDelta_{\omega 0}^{(m)}(r, v_{\text{out}})] . \qquad (15.24)$$

It would not be reasonable and, perhaps, not even meaningful to extend the double sum in Eq. (15.23) actually up to infinity. Since the order of magnitude of the various contributions to $\varPhi_1$ is expected to be roughly given by the sum, $l + m$, the summation in Eq. (15.23) shall be extended up to a certain finite value of $l + m$.

Two simple statements regarding $\varPhi_1(l, m)$ can easily be made. It is

$$\varPhi_1(l, 0) = 0 \qquad (15.25)$$

and[*]

$$\varPhi_1(l, m) = \varPhi_1^*(m, l) - 8\pi^4 N \oint d\Omega \oint d^2 r \cdot v T \sum_\omega \varDelta_{\omega 0}^{(l)}(r, -v) \varDelta_{\omega 0}^{(m)}(r, v) . \qquad (15.26)$$

Eq. (15.25) follows immediately from Eq. (15.24) since $\varDelta_{\omega 0}^{(0)}(r, v)$ does not depend upon the direction of $v$. Eq. (15.26) is derived if use is made of the general properties of the reflexion kernel, Eqs. (10.18) through (10.20). The second term on the righthand side of Eq. (15.26) is easily evaluated with the help of Eqs. (13.6), (13.48), and generalizations thereof. The integral vanishes if the sum, $l + m$, is an even number. More important is that these integrals may be combined with $\varPhi_0$ in such a way that the combination is evidently real. Since also the remaining contributions to $\varPhi_1$ are real, $\varPhi$ itself is real; so $\varDelta(r)$ and $\varDelta^*(r)$ may be varied independently. The generalized GL equation results if $\varPhi_0$ is varied with respect to $\varDelta^*(r)$. Low order contribution, to the boundary conditions are obtained by varying $\varPhi_1$ in the same way.

---

[*] From now on the dash on the summation sign, which actually is of no relevance, shall be disregarded.

It follows from Eq. (15.25) that it is

$$\Phi_1(0, 0) = \Phi_1(1, 0) = 0 \tag{15.27}$$

whereas Eq. (15.26) immediately leads to

$$\Phi_1(0, 1) = \frac{7\zeta(3)}{12} N\xi_0^2(T) \oint \Delta^*(r) \, d^2 r \cdot \tilde{\partial}\Delta(r). \tag{15.28}$$

The surface integral in Eq. (15.13) is, for $n = 0$, identical with Eq. (15.28). If we do not go beyond $l + m = 1$, variation of Eq. (15.28) with respect to $\Delta^*(r)$ again leads to the boundary condition (15.1).

The only non-vanishing contribution to $\Phi_1$ with $l + m = 2$ is given by

$$\Phi_1(1, 1) = -\frac{15\zeta(4)}{128} N\xi_0^3(T) \oint |d^2 r| \, p \, |n \times \tilde{\partial}\Delta(r)|^2. \tag{15.29}$$

A term containing the surface-normal gauge invariant gradient of the pair potential was neglected on the righthand side of Eq. (15.29) because of Eq. (15.1), which is still approximately true. This further term need not be taken into account before going up to $l + m = 4$. The parameter $p$ in Eq. (15.29), which may be position dependent, is the diffuseness defined in Eq. (10.24). A more general definition will be given in Eq. (15.42). In this more general case, Eq. (10.25) is replaced by Eq. (15.44). The proof of Eq. (15.29) will be given below in small print.

The collection of terms with $l + m = 3$ leads to

$$\Phi_1(3, 0) + \Phi_1(2, 1) + \Phi_1(1, 2) + \Phi_1(0, 3)$$
$$= \frac{31\zeta(5)}{240} N\xi_0^4(T) \left[ 2ie \oint (d^2 r \times H(r)) \right.$$
$$\cdot \left\{ \left(3 + \frac{p'}{2}\right) \Delta^*(r) \tilde{\partial}\Delta(r) + \left(1 - \frac{p'}{2}\right) (\tilde{\partial}\Delta(r))^* \Delta(r) \right\} \tag{15.30}$$
$$+ 2 \oint |d^2 r| \frac{\partial n_k}{\partial x_l} \left\{ \left(1 - \frac{p'}{2}\right) (\tilde{\partial}_l \Delta(r))^* \tilde{\partial}_k \Delta(r) - \frac{p'}{2} (\tilde{\partial}_k \Delta(r))^* \tilde{\partial}_l \Delta(r) \right\} \right].$$

Also these contributions to $\Phi_1$ will be derived below in small print. It is $p' = p$ for the family (10.24) of reflexion kernels. The general definition of $p'$ will be given in Eq. (15.54). In the second integral, which vanishes on plane surfaces, $n_k$ is a Cartesian component of the surface normal unit vector pointing towards the exterior. Some contributions to the righthand side of Eq. (15.30) have not been written down. They do not contribute before we go beyond $l + m = 3$.

It is possible to work directly with the contributions to $\Phi_1$ given in Eqs. (15.28), (15.29), and (15.30). In the spirit of the idea of extrapolating

9*

the solution to the (generalized) GL equation through the surface layer, $\Phi_1$ has to be made stationary within the manyfold of solutions to this equation (for which $\Phi_0$ vanishes). If the integrand in $\Phi_1$ can be written in such a way that $\Delta^*(r)$ appears without derivatives, it is also possible to obtain directly local boundary conditions by varying $\Phi_1$ with respect to $\Delta^*(r)$. This cannot be done for arbitrarily high order contributions to $\Phi_1$, but it can for the contributions written down. Since only surface parallel derivatives of $\Delta^*(r)$ occur, it is possible to perform integrations by parts according to the rule

$$\oint d^2 r \times (\tilde{\partial} f(r))^* \, g(r) = - \oint f^*(r) \, d^2 r \times \tilde{\partial} g(r). \qquad (15.31)$$

The local boundary condition obtained in this way is given by

$$n \cdot \tilde{\partial} \, \Delta(r) = - \frac{45 \, \zeta(4)}{224 \, \zeta(3)} \, \xi_0(T) \, [p(n \times \tilde{\partial}) \cdot (n \times \tilde{\partial}) \, \Delta(r) + (\operatorname{grad} p) \cdot \tilde{\partial} \, \Delta(r)]$$

$$- \frac{31 \, \zeta(5)}{140 \, \zeta(3)} \, \xi_0^2(T) \left[ 2 i e (n \times H(r)) \cdot \left[ (2 + p') \, \tilde{\partial} \, \Delta(r) + \frac{1}{2} (\operatorname{grad} p') \, \Delta(r) \right] \right.$$

$$\left. + \varepsilon_{lmn} \varepsilon_{lkr} n_k \tilde{\partial}_r \left( \frac{\partial n_i}{\partial x_m} (2 - p') - \frac{\partial n_m}{\partial x_i} p' \right) n_n \tilde{\partial}_i \Delta(r) \right] + \cdots. \qquad (15.32)$$

All terms on the righthand side are small in the sense of the GL approximation. If they are put equal to zero, the standard boundary condition (15.1) is again obtained. $\varepsilon_{lmn}$ is the totally antisymmetric Ricci tensor with $\varepsilon_{123} = +1$. It is

$$\varepsilon_{lmn} \varepsilon_{lkr} = \delta_{mk} \delta_{nr} - \delta_{mr} \delta_{nk}. \qquad (15.33)$$

Eventual contributions from edges were not written down in Eq. (15.31). Edges in any case pose a difficult problem since a radius of curvature becomes equal to zero on edges and is no longer big as compared to $\xi_0(T)$. This assumption was, however, basic for our analysis since it lead to Eq. (15.19).

The task of solving special problems with the help of the boundary conditions will be taken up in Chapter III.

Now the derivation of the various contributions to $\Phi_1$ shall be supplied. Regarding first the case $l + m = 2$, it is easily seen with the help of Eqs. (15.25) and (15.26) that it is

$$\Phi_1(2, 0) = \Phi_1(0, 2) = 0. \qquad (15.34)$$

The only non-vanishing contribution is given by

$$\Phi_1(1, 1) = - \frac{N}{2} \, T \sum_\omega \frac{1}{(2|\omega|)^4} \oint |d^2 r| \, \langle (v_{\text{out}} \cdot \tilde{\partial} \Delta(r))^* \, (v_{\text{out}} - v_{\text{in}}) \cdot \tilde{\partial} \Delta(r) \rangle. \qquad (15.35)$$

Here, the notation

$$\langle f(v_{\text{out}}, v_{\text{in}}) \rangle := - \int d\Omega_{\text{out}} d\Omega_{\text{in}} (n \cdot v_{\text{out}}) \, R(r; v_{\text{out}}, v_{\text{in}}) \, f(v_{\text{out}}, v_{\text{in}}) \qquad (15.36)$$

has been used. The surface normal unit vector, $n$, points towards the exterior. The weight factor, $-n \cdot v_{out} R(r; v_{out}, v_{in})$, is a non-negative quantity as a consequence of Eq. (10.17). Further it is

$$\langle f(v_{out}, v_{in}) \rangle = \langle f(-v_{in}, -v_{out}) \rangle \qquad (15.37)$$

because of Eq. (10.19). For the further evaluation it shall be assumed that the reflexion kernel, $R(r; v_{out}, v_{in})$, is invariant under rotations about $n$. Then it is

$$\langle (v_{out} \cdot \tilde{\partial} \Delta(r))^* (v_{out} - v_{in}) \cdot \tilde{\partial} \Delta(r) \rangle = \langle (v_{out} \cdot n)(v_{out} - v_{in}) \cdot n \rangle |n \cdot \tilde{\partial} \Delta(r)|^2 \\ + \tfrac{1}{2} \langle (v_{out} \times n) \cdot [(v_{out} - v_{in}) \times n] \rangle |n \times \tilde{\partial} \Delta(r)|^2 . \qquad (15.38)$$

The first term on the righthand side may be neglected as long as we do not go beyond $l + m = 3$ since Eq. (15.1) is still approximately true; cf. also Eq. (15.32).

It is immediately seen that $\langle (v_{out} \times n) \cdot [(v_{out} - v_{in}) \times n] \rangle$ vanishes for specular reflexion since, in this case, it is

$$(v_{out} - v_{in}) \times n = 0 , \qquad (15.39)$$

whenever the reflexion kernel is different from zero. For diffuse reflexion, $v_{out}$ and $v_{in}$ are uncorrelated so that it is

$$\langle (v_{out} \times n) \cdot (v_{in} \times n) \rangle \langle 1 \rangle = \langle v_{out} \times n \rangle \cdot \langle v_{in} \times n \rangle . \qquad (15.40)$$

Each factor on the righthand side vanishes for rotational invariance, about $n$, of the reflexion kernel. We are left with

$$\langle (v_{out} \times n) \cdot [(v_{out} - v_{in}) \times n] \rangle = \langle (v_{out} \times n)^2 \rangle = \frac{\pi}{2} v^3 , \qquad (15.41)$$

which is valid for diffuse reflexion. The last equality sign is, however, correct for arbitrary reflexion kernels. This is seen if use is made of Eq. (10.20). It follows from a combination of the results for specular and diffuse reflexion that it is

$$\langle (v_{out} \times n) \cdot [(v_{out} - v_{in}) \times n] \rangle = : p \frac{\pi}{2} v^3 \qquad (15.42)$$

for the one-parametric family (10.24) of reflexion kernels.

For arbitrary, rotationally invariant reflexion kernels, Eq. (15.42) may be regarded as the definition of $p$. The Cauchy inequality

$$\langle (v_{out} \times n) \cdot (v_{in} \times n) \rangle^2 \leqq \langle (v_{out} \times n)^2 \rangle \langle (v_{in} \times n)^2 \rangle = \langle (v_{out} \times n)^2 \rangle^2 = \left( \frac{\pi}{2} v^3 \right)^2 , \qquad (15.43)$$

where use was also made of Eq. (15.37), leads to

$$0 \leqq p \leqq 2 . \qquad (15.44)$$

$p = 0$ still means specular reflexion. Values of $p$ bigger than one seem unphysical since they would mean that the surface parallel velocity component is, on the average, reversed in the reflexion process. Combining Eqs. (15.35), (15.38), and (15.42) we obtain Eq. (15.29).

Turning now to $l + m = 3$, we see from Eqs. (15.25) and (15.26) that it is

$$\Phi_1(3, 0) = 0 \qquad (15.45)$$

and

$$\Phi_1(0, 3) = \frac{2\pi}{15} T \sum_\omega \frac{1}{(2|\omega|)^5} N v^4 \delta_{jklm} \oint |d^2 r| \, \Delta^*(r) \, n_j \tilde{\partial}_k \tilde{\partial}_l \tilde{\partial}_m \Delta(r) . \qquad (15.46)$$

The angular integral was already evaluated with the help of Eqs. (13.48) and (13.49). It is

$$(\boldsymbol{n} \cdot \tilde{\partial})(\tilde{\partial} \cdot \tilde{\partial}) \, \Delta(\boldsymbol{r}) \approx 0 \tag{15.47}$$

(and equal to zero up to the present approximation). Eq. (15.47) follows from Eqs. (13.1) and (15.1); it is not necessary to apply here a generalized GL approximation as long as it is not intended to go beyond $l + m = 3$. Not in all terms on the righthand side of Eq. (15.46), the gradient operators have the correct order so that Eq. (15.47) is applicable. In all other cases the commutation relation (11.57) is applied. If use is also made of

$$\operatorname{curl} \boldsymbol{H}(\boldsymbol{r}) = 0, \tag{15.48}$$

we finally obtain

$$\delta_{jklm} \oint |\mathrm{d}^2 r| \, \Delta^*(\boldsymbol{r}) \, n_j \tilde{\partial}_k \tilde{\partial}_l \tilde{\partial}_m \Delta(\boldsymbol{r}) = 3(2ie) \oint \Delta^*(\boldsymbol{r}) \, (\mathrm{d}^2 \boldsymbol{r} \times \boldsymbol{H}(\boldsymbol{r})) \cdot \tilde{\partial} \Delta(\boldsymbol{r}). \tag{15.49}$$

From Eqs. (15.24) and (15.22) we obtain

$$\Phi_1(2, 1) = -\frac{N}{2} T \sum_\omega \frac{1}{(2|\omega|)^5} \oint |\mathrm{d}^2 r| \, \langle [(\boldsymbol{v}_{\mathrm{out}} \cdot \tilde{\partial})^2 \, \Delta(\boldsymbol{r})]^* \, (\boldsymbol{v}_{\mathrm{out}} - \boldsymbol{v}_{\mathrm{in}}) \cdot \tilde{\partial} \Delta(\boldsymbol{r}) \rangle. \tag{15.50}$$

This equation is formulated with the help of a notation introduced in Eq. (15.36). For a reflexion kernel which is rotationally invariant about $\boldsymbol{n}$, we get

$$\langle [(\boldsymbol{v}_{\mathrm{out}} \cdot \tilde{\partial})^2 \, \Delta(\boldsymbol{r})]^* \, (\boldsymbol{v}_{\mathrm{out}} - \boldsymbol{v}_{\mathrm{in}}) \cdot \tilde{\partial} \Delta(\boldsymbol{r}) \rangle \tag{15.51}$$
$$= \tfrac{1}{2} \langle (\boldsymbol{v}_{\mathrm{out}} \cdot \boldsymbol{n}) \, (\boldsymbol{v}_{\mathrm{out}} \times \boldsymbol{n}) \cdot [(\boldsymbol{v}_{\mathrm{out}} - \boldsymbol{v}_{\mathrm{in}}) \times \boldsymbol{n}] \rangle \, \varepsilon_{jkl} \varepsilon_{jmn} n_p n_k n_m [(\tilde{\partial}_p \tilde{\partial}_l + \tilde{\partial}_l \tilde{\partial}_p) \, \Delta(\boldsymbol{r})]^* \, \tilde{\partial}_n \Delta(\boldsymbol{r}) + \cdots.$$

in close analogy to Eq. (15.38). The dots denote terms which may be discarded because of the approximate validity of Eq. (15.1). The Ricci tensor was explained in connection with Eq. (15.32). If Eq. (15.33) is applied to the product of two such tensors, only one term on the righthand side contributes because of the approximate validity of Eq. (15.1). In this way we obtain

$$\varepsilon_{jkl} \varepsilon_{jmn} n_p n_k n_m [(\tilde{\partial}_p \tilde{\partial}_l + \tilde{\partial}_l \tilde{\partial}_p) \, \Delta(\boldsymbol{r})]^* \, \tilde{\partial}_n \Delta(\boldsymbol{r}) = n_p [(\tilde{\partial}_p \tilde{\partial}_n + \tilde{\partial}_n \tilde{\partial}_p) \, \Delta(\boldsymbol{r})]^* \, \tilde{\partial}_n \Delta(\boldsymbol{r}) + \cdots \tag{15.52}$$

where the dots again denote a term to be discarded. Now we aim at bringing $n_p \tilde{\partial}_p$ directly in front of $\Delta(\boldsymbol{r})$ and making again use of the approximate validity of Eq. (15.1). Applying also the commutation relation (11.57), we finally obtain

$$n_p [(\tilde{\partial}_p \tilde{\partial}_n + \tilde{\partial}_n \tilde{\partial}_p) \, \Delta(\boldsymbol{r})]^* \, \tilde{\partial}_n \Delta(\boldsymbol{r})$$
$$\tag{15.53}$$
$$= 2ie \Delta^*(\boldsymbol{r}) \, (\boldsymbol{n} \times \boldsymbol{H}(\boldsymbol{r})) \cdot \tilde{\partial} \Delta(\boldsymbol{r}) - 2 \frac{\partial n_p}{\partial x_n} (\tilde{\partial}_p \Delta(\boldsymbol{r}))^* \, \tilde{\partial}_n \Delta(\boldsymbol{r}) + \cdots.$$

The second term on the righthand side vanishes on a plane surface.

The other factor on the righthand side of Eq. (15.51) is written in the form

$$\langle (\boldsymbol{v}_{\mathrm{out}} \cdot \boldsymbol{n}) \, (\boldsymbol{v}_{\mathrm{out}} \times \boldsymbol{n}) \cdot [(\boldsymbol{v}_{\mathrm{out}} - \boldsymbol{v}_{\mathrm{in}}) \times \boldsymbol{n}] \rangle = : -p' \frac{4\pi}{15} v^4 \tag{15.54}$$

which relation defines the parameter $p'$. It is easily seen that it is $p' = 0$ for specular reflexion [cf. Eq. (15.39)] and $p' = 1$ for diffuse reflexion. Therefore, it is $p' = p$ for the family (10.24) of reflexion kernels. From the Cauchy inequality

$$\langle (\boldsymbol{v}_{\mathrm{out}} \cdot \boldsymbol{n}) \, (\boldsymbol{v}_{\mathrm{out}} \times \boldsymbol{v}) \cdot (\boldsymbol{v}_{\mathrm{in}} \times \boldsymbol{n}) \rangle^2 \leqq \langle (\boldsymbol{v}_{\mathrm{out}} \cdot \boldsymbol{n})^2 \, (\boldsymbol{v}_{\mathrm{out}} \times \boldsymbol{n})^2 \rangle \, \langle (\boldsymbol{v}_{\mathrm{in}} \times \boldsymbol{n})^2 \rangle = \frac{\pi}{6} v^5 \frac{\pi}{2} v^3, \tag{15.55}$$

it follows that it is

$$1 - \frac{5}{8}\sqrt{3} \leqq p' \leqq 1 + \frac{5}{8}\sqrt{3}. \tag{15.56}$$

Values of $p'$ bigger than one seem again unplausible.

The last still missing contribution is obtained from Eq. (15.26),

$$\Phi_1(1, 2) = \Phi_1^*(2, 1) - \frac{2\pi}{15} T \sum_\omega \frac{1}{(2|\omega|)^5} N v^4 \delta_{jklm} \oint |\mathrm{d}^2 r \, |n_j(\tilde{\partial}_k \Delta(r))^* \, \tilde{\partial}_l \tilde{\partial}_m \Delta(r). \tag{15.57}$$

If the integrand is rearranged in the same way as in Eq. (15.53), we obtain

$$\delta_{jklm} n_j(\tilde{\partial}_k \Delta(r))^* \, \tilde{\partial}_l \tilde{\partial}_m \Delta(r) = -2ie(n \times H(r)) \cdot (\tilde{\partial}\Delta(r))^* \, \Delta(r)$$

$$-2 \frac{\partial n_k}{\partial x_l} (\tilde{\partial}_l \Delta(r))^* \, \tilde{\partial}_k \Delta(r) + \cdots. \tag{15.58}$$

### c) Interfaces (for Clean Metals only)

If a sample consists of several metals in electric contact, also boundary conditions on the interfaces between different metals are required. They shall now be derived [44] though the problem of metals in electric contact is, in the GL approximation, not very interesting from a physical point of view. The reason is that the temperature has in any case to be close to the transition temperatures of each of the metals in order that the GL approximation is valid at all. Therefore, the transition temperatures of the various metals in electric contact have to be close to each other so that no drastic phenomena can be expected.

Whereas the lowest order boundary condition (15.1) on external surfaces could be derived for arbitrary impurity concentration (Section 15a) this has so far not been possible for the lowest order boundary conditions on interfaces. They were only obtained for clean metals ($n = 0$). We shall not go beyond the lowest order boundary conditions.

The general ideas of the derivation are the same as in the case of surfaces (Section 15b). We again start from the functional (15.5). The spatial integration is now extended over the whole sample consisting of several metals. Coupling constant, $g$, and density of states, $N$, will in general be different in different metals but they are still constant inside each metal. Eq. (15.6) is again used. The somewhat complicated cut-off procedure proposed in Eq. (10.2) does not come into play here since $\Delta_{\omega 0}(r, v)$ depends only upon the pair potential and its low order derivatives in the same point $r$ and since the differential equation (15.7) for $\Delta_{\omega 1}(r, v)$ does not contain $\Delta(r)$ at all. Therefore, the cut-off in Eq. (15.5) has to be made at the Debye frequency of that metal in which the point $r$ is situated. In a less formal way we might say that the cut-off procedure of Eq. (10.2) would be relevant only in the interface layer through which actually a simple extrapolation of the pair potential is made.

Also Eq. (15.15) may still be used. $\Phi_0$ as given in Eq. (15.16) is now represented by an integral over the whole sample which consists of several metals. $\Phi_1$ [Eq. (15.17)] is a sum of boundary integrals over the complete boundary, $B_j$, of each individual sample,

$$\Phi_1 = 8\pi^4 \sum_j N_j \oint d\Omega \oint_{B_j} d^2r \cdot v \, T \sum_\omega \Delta^*_{\omega 0}(r, -v_j) \Delta_{\omega 1}(r, v_j). \quad (15.59)$$

$N_j$ is the density of states of the metal $j$. The distribution function $\Delta_{\omega 1}(r, v_j)$ still satisfies the homogeneous Boltzmann equation (15.7), but the inhomogeneous boundary condition (15.8) has to be replaced by more complicated conditions following from Eq. (10.26). As in that equation, we shall from now on assume that the sample consists of exactly two metals, number 1 and number 2.

If both metals are clean and if the linear dimensions of the homogeneous parts of the sample are big as compared to the corresponding length $\xi_0(T)$, essentially Eq. (15.19) is again valid. We now write it in the form

$$\Delta_{\omega 1}(r, v_{1\,\text{in}}) = 0 = \Delta_{\omega 1}(r, v_{2\,\text{in}}). \quad (15.60)$$

Instead of Eq. (15.20) it is

$$\Delta_{\omega 1}(r, v_{1\,\text{out}}) = - \Delta_{\omega 0}(r, v_{1\,\text{out}}) + \int R_1(r; v_{1\,\text{out}}, v_{1\,\text{in}}) \Delta_{\omega 0}(r, v_{1\,\text{in}}) d\Omega_{1\,\text{in}}$$
$$+ \int T_1(r; v_{1\,\text{out}}, v_{2\,\text{in}}) \Delta_{\omega 0}(r, v_{2\,\text{in}}) d\Omega_{2\,\text{in}}. \quad (15.61)$$

There holds a second equation with subscripts 1 and 2 interchanged. Eqs. (15.60) and (15.61), and this second equation are then inserted into Eq. (15.59). The righthand side is in this way expressed by the distribution functions $\Delta_{\omega 0}(r, v_1)$ and $\Delta_{\omega 0}(r, v_2)$ which are known. Again the expansion (15.23) can be made. We do not go beyond $l + m = 1$ since only lowest order boundary conditions shall be derived. The details of the calculation will be presented below in small print.

The result is given by the boundary conditions

$$\frac{9\zeta(2) N_1 \xi_{10}(T) t_1}{7\zeta(3)(2 - t'_1 - t'_2)} (\Delta_2(r) - \Delta_1(r)) \quad (15.62)$$
$$= N_1 \xi^2_{10}(T) \, n \cdot \tilde{\partial} \Delta_1(r) = N_2 \xi^2_{20}(T) \, n \cdot \tilde{\partial} \Delta_2(r).$$

Here, the interface normal unit vector, $n$, points from the interface into metal 2. $\Delta_1(r)$ and $\Delta_2(r)$ are boundary values of the pair pontential in the metal 1 and 2 respectively. The pair potential satisfies the appropriate GL equation (13.1) in each metal. $N_1$ and $N_2$ are the densities of states in the respective metals. The lengthes $\xi_{10}(T)$ and $\xi_{20}(T)$ are defined by

$$\xi_{10}(T) := \frac{v_1}{2\pi T}, \qquad \xi_{20}(T) := \frac{v_2}{2\pi T} \quad (15.63)$$

with $v_1$ and $v_2$ being the Fermi velocities in the respective metal; cf. Eq. (13.11). The coefficients $t_1$, $t_2$, $t'_1$, and $t'_2$ depend upon the transmission properties of the interface. They are defined in Eqs. (15.73), (15.75), and (15.77), respectively. They satisfy certain inequalities which will also be given below. As a consequence, the factor of $\Delta_2(r) - \Delta_1(r)$ in Eq. (15.62) is a non-negative quantity. It further is

$$N_1 \xi_{10}(T)\, t_1 = N_2 \xi_{20}(T)\, t_2 \,. \tag{15.64}$$

The coefficients may of course be position dependent on the surface. It is assumed in the derivation of Eq. (15.62) that reflexion and transmission kernels are rotationally invariant about $n$.

On an isolating interface with vanishing transmission kernels it is

$$t_1 = t'_1 = t'_2 = 0 \,. \tag{15.65}$$

The boundary conditions (15.62) then reduce to boundary condition (15.1) on external surfaces as had to be expected. It is, on the other hand, possible to have

$$t'_1 + t'_2 = 2 \tag{15.66}$$

provided that Eq. (6.25) holds. In this case, the boundary conditions (15.62) split into

$$\begin{aligned}
\Delta_1(r) &= \Delta_2(r) \,, \\
N_1 \xi_{10}^2(T)\, n \cdot \tilde{\partial}\, \Delta_1(r) &= N_2 \xi_{20}^2(T)\, n \cdot \tilde{\partial}\, \Delta_2(r) \,.
\end{aligned} \tag{15.67}$$

Boundary conditions of this type have earlier been obtained by *Zaitsev* [20]. Since he assumes free electrons and equal Fermi velocities in the adjoining metals, the second Eq. (15.67) becomes

$$n \cdot \tilde{\partial}\, \Delta_1(r) = n \cdot \tilde{\partial}\, \Delta_2(r) \,. \tag{15.68}$$

Since the boundary conditions (15.67) do not contain any parameter characterizing the interface, it is recommendable to use them also when Eq. (6.25) does not hold. In this case, we shall refer to Eq. (15.67) as to simplified boundary conditions. It is hoped that they will lead to reasonable results provided that the electric contact through the interface is sufficiently good.

If metal 2 is a normal metal in the sense of the somewhat academic condition

$$g_2 = 0 \,, \tag{15.69}$$

the pair potential in metal 2 vanishes as a consequence of the gap equation (10.1). The boundary condition for the pair potential in metal 1 is then given by

$$\xi_{10}(T)\, \boldsymbol{n} \cdot \tilde{\partial}\, \varDelta_1(\boldsymbol{r}) = -\frac{9\, \zeta(2)\, t_1}{7\, \zeta(3)\, (2 - t_1' - t_2')}\, \varDelta_1(\boldsymbol{r})\,. \qquad (15.70)$$

The simplified boundary condition reads

$$\varDelta_1(\boldsymbol{r}) = 0 \qquad (15.71)$$

in this case.

The derivation of the boundary conditions (15.62) shall be given in a rather compact form since it proceeds in essentially the same manner as in Section 15b. First we derive the various contributions to the functional $\Phi_1$ [Eq. (15.59)]. Decisive use is made of Eqs. (15.60) and (15.61). The general properties (10.27) through (10.31) will be applied in the following analysis. We again make the expansion (15.23). Eq. (15.24) is, of course, replaced by a more complicated relation.

The interface contribution to $\Phi_1(0, 0)$ is given by

$$\Phi_1(0, 0) = \frac{3\zeta(2)}{8}\, N_1 \xi_{10}(T) \int |d^2 r|\, t_1\, |\varDelta_1(\boldsymbol{r}) - \varDelta_2(\boldsymbol{r})|^2 \qquad (15.72)$$

with $t_1$ defined by

$$\pi t_1 v_1 := -\int d\Omega_{1\,\text{out}}\, d\Omega_{2\,\text{in}}\, \boldsymbol{n} \cdot \boldsymbol{v}_{1\,\text{out}}\, T_1(\boldsymbol{r};\, \boldsymbol{v}_{1\,\text{out}},\, \boldsymbol{v}_{2\,\text{in}})\,. \qquad (15.73)$$

Here, the normal unit vector, $\boldsymbol{n}$, points into metal 2. $t_1$ obeys the inequality

$$0 \leqq t_1 \leqq 1\,. \qquad (15.74)$$

which is derived with the help of Eqs. (10.27) and (10.31). Introducing $t_2$ in the same way by

$$\pi t_2 v_2 := \int d\Omega_{2\,\text{out}}\, d\Omega_{1\,\text{in}}\, \boldsymbol{n} \cdot \boldsymbol{v}_{2\,\text{out}}\, T_2(\boldsymbol{r};\, \boldsymbol{v}_{2\,\text{out}},\, \boldsymbol{v}_{1\,\text{in}})\,, \qquad (15.75)$$

we derive Eq. (15.64) from Eq. (10.30).

For reflexion and transmission kernels which are invariant under rotations about $\boldsymbol{n}$, the interface contribution to $\Phi_1(1, 0)$ is given by

$$\Phi_1(1, 0) = -\frac{7\zeta(3)}{24} \int d^2 r \cdot [N_1 \xi_{10}^2(T)\, t_1' \tilde{\partial} \varDelta_1(\boldsymbol{r}) + N_2 \xi_{20}^2(T)\, t_2' \tilde{\partial} \varDelta_2(\boldsymbol{r})]^* (\varDelta_1(\boldsymbol{r}) - \varDelta_2(\boldsymbol{r}))\,.$$
$$(15.76)$$

The coefficient $t_1'$ is defined by

$$\frac{2\pi}{3}\, t_1' v_1^2 := \int d\Omega_{1\,\text{out}}\, d\Omega_{2\,\text{in}}\, (\boldsymbol{n} \cdot \boldsymbol{v}_{1\,\text{out}})^2\, T_1(\boldsymbol{r};\, \boldsymbol{v}_{1\,\text{out}},\, \boldsymbol{v}_{2\,\text{in}})\,. \qquad (15.77)$$

$t_2'$ is defined in an analogous manner. It is

$$0 \leqq t_1' \leqq 1\,. \qquad (15.78)$$

It further is

$$\tfrac{3}{4} t_1^2 \leqq t_1' \leqq \tfrac{3}{2} t_1 \qquad (15.79)$$

where the second inequality sign follows from

$$0 \leqq -\boldsymbol{n} \cdot \boldsymbol{v}_{1\,\text{out}} \leqq 1 \qquad (15.80)$$

and the first one from an appropriate Cauchy inequality.

If

$$\int R_1(r; v_{1\,\text{out}}, v_{1\,\text{in}})\, d\Omega_{1\,\text{in}} = r_1 \tag{15.81}$$

is independent of $v_{1\,\text{out}}$, then it is [13]

$$t_1 = t_1' = 1 - r_1. \tag{15.82}$$

Eq. (15.81) is true for specular reflexion,

$$R_1(v_{1\,\text{out}}, v_{1\,\text{in}}) = r_1\, \hat{\delta}(v_{1\,\text{out}}, v_{1\,\text{in}} - 2\,n(n \cdot v_{1\,\text{in}})), \tag{15.83}$$

for diffuse reflexion,

$$R_1(v_{1\,\text{out}}, v_{1\,\text{in}}) = r_1\, \frac{n \cdot v_{1\,\text{in}}}{\pi v_1}, \tag{15.84}$$

and for any linear combination of the two kernels with non-negative weight factors of sum equal to one.

In the same manner we obtain

$$\Phi_1(0, 1) = \Phi_1^*(1, 0)$$
$$+ \frac{7\zeta(3)}{12} \int d^2r \cdot [N_1\,\xi_{10}^2(T)\,\Delta_1^*(r) - N_2\,\xi_{20}^2(T)\,\Delta_2^*(r)]\, \tilde{\partial}\Delta_2(r). \tag{15.85}$$

This equation is of a similar structure as Eq. (15.26).

The interface contribution to $\Phi_1$ is, up to the desired accuracy, given by the sum of the righthand sides of Eqs. (15.72), (15.76), and (15.85). Because of the integral on the right-hand side of Eq. (15.85), the interface contribution to $\Phi_1$ is real within the manifold of solutions to the GL equations. It is, therefore, again permitted to only vary $\Delta_1^*(r)$ and $\Delta_2^*(r)$ in order to derive the explicit boundary conditions. Putting the factor of $\delta\Delta_1^*(r)$ equal to zero, we obtain

$$9\,\zeta(2)\,N_1\,\xi_{10}(T)\,t_1(\Delta_1(r) - \Delta_2(r)) - 7\,\zeta(3)\,[N_1\,\xi_{10}^2(T)\,t_1'\,n \cdot \tilde{\partial}\Delta_1(r)$$
$$+ N_2\,\xi_{20}^2(T)\,t_2'\,\tilde{\partial}\Delta_2(r)] + 14\,N_1\,\xi_{10}^2(T)\,n \cdot \tilde{\partial}\Delta_1(r) = 0. \tag{15.86}$$

Putting the factor of $\delta\Delta_2^*(r)$ equal to zero, we obtain an analogous expression with subscripts 1 and 2 interchanged. The boundary conditions (15.62) follows from these two equations if also Eq. (15.64) is used.

There are, however, also interface contributions proportional to $n \cdot [\tilde{\partial}\delta\Delta_j(r)]^*$ $(j = 1, 2)$. These variations cannot be chosen independently of the $\delta\Delta_j^*(r)$ since $\Delta_j(r)$ itself obeys the GL equation (variation within the manifold of solutions to this equation!). But $n \cdot [\tilde{\partial}\delta\Delta_j(r)]^*$ is small as compared to $\delta\Delta_j^*(r)$ in the sense of the GL approximation, and it is multiplied by $\Delta_1(r) - \Delta_2(r)$ which is also small in the sense of this approximation because of Eq. (15.86). These contributions have, therefore, to be neglected altogether.

## d) A Variational Principle

For practical calculations it is always helpful to have a variational principle available. The variational principle based upon the functional (15.5) is, however, not practical in the case of metals in contact because of the derivative terms occurring in $\Phi_1$. A rather sophisticated argument was applied at the end of Section 15c in order to get rid of unpleasant consequences of these terms. It is, therefore, recommendable to construct the variational principle directly from the differential equations and the boundary conditions.

We again regard a sample consisting of two clean metals, number 1 and number 2, in order to simplify the notation. The GL equation is given by

$$\tilde{\partial} \cdot \tilde{\partial} \, \Delta_j(r) = - \lambda_j(T) \, \Delta_j(r) \qquad (15.87)$$

with $j$ equal to 1 or 2. $\lambda_j(T)$ is, according to Eq. (13.13), given by

$$\lambda_j(T) = \frac{12}{7 \, \zeta(3) \, \xi_{j0}^2(T_{cj})} \, \frac{T_{cj} - T}{T_{cj}} . \qquad (15.88)$$

The lengthes $\xi_{j0}(T)$ were defined in Eq. (15.63). $T_{cj}$ is the transition temperature of the metal in question. The reader is reminded that $T_{c1}$ and $T_{c2}$ have to be close to each other in order that the GL approximation is valid at all. With the notation

$$A_{12}(r) := \frac{9 \, \zeta(2) \, N \, \xi_{10}(T_{c1}) \, t_1(r)}{7 \, \zeta(3) \, (2 - t_1'(r) - t_2'(r))} \geqq 0, \qquad (15.89)$$

the boundary condition (15.62) may be written as

$$A_{12}(r) \, (\Delta_2(r) - \Delta_1(r)) = N_1 \xi_{10}^2(T_{c1}) \, \boldsymbol{n} \cdot \tilde{\partial} \, \Delta_1(r) = N_2 \xi_{20}^2(T_{c2}) \, \boldsymbol{n} \cdot \tilde{\partial} \, \Delta_2(r) . \qquad (15.90)$$

Now we regard the functional,

$$\Psi = \sum_j N_j \xi_{0j}^2(T_{cj}) \int_{V_j} (|\tilde{\partial} \, \Delta_j(r)|^2 - \lambda_j(T) |\Delta_j(r)|^2) \, \mathrm{d}^3 r \qquad (15.91)$$
$$+ \int A_{12}(r) \, |\Delta_2(r) - \Delta_1(r)|^2 \, |\mathrm{d}^2 r| ,$$

of the pair potential $\Delta_j(r)$. The volume integrals are extended over the respective metals. The surface integral is extended over the interface between the two metals. $\Psi$ is a real functional. Putting the variation with respect to $\Delta_j^*(r)$ equal to zero, we obtain both the GL equations (15.87) and the boundary conditions (15.90). The simplified boundary conditions (15.67) may be formulated in terms of a similar variational principle. In this case we put

$$\Psi = \sum_j N_j \xi_{0j}^2(T_{cj}) \int_{V_j} (|\tilde{\partial} \, \Delta_j(r)|^2 - \lambda_j(T) |\Delta_j(r)|^2) \, \mathrm{d}^3 r \qquad (15.92)$$

without a surface term and postulate the subsidiary condition

$$\Delta_1(r) = \Delta_2(r) \qquad (15.93)$$

on the interface.

## 16. Diffusion Approximation: Boundary Conditions

The diffusion approximation (Section 14) is valid at all temperatures provided that the electronic mean free path is sufficiently short [dirty limit, Eq. (14.15)]. If the sample is bounded by an external surface or if it consists of several metals in electric contact, the diffusion equation (14.8) or (14.10) has to be supplemented by boundary conditions which shall be derived in this section.

The derivation is again very simple for a specularly reflecting external surface. If the considerations made in Section 15 in connection with the distribution function (15.2) are now repeated for the distribution function (14.1), we derive the boundary condition

$$\boldsymbol{n} \cdot \boldsymbol{H}_\omega(\boldsymbol{r}) = 0 , \tag{16.1}$$

where $\boldsymbol{n}$ is again the normal unit vector in the surface point $\boldsymbol{r}$. Application of Eq. (14.6) leads to

$$\boldsymbol{n} \cdot \tilde{\partial} \, H_\omega(\boldsymbol{r}) = 0 \tag{16.2}$$

which is the boundary condition actually to be used in connection with Eq. (14.8). The analogous boundary condition to be applied in conjunction with Eq. (14.10) is given by

$$\boldsymbol{n} \cdot \tilde{\partial} \, K_\omega(\boldsymbol{r}, \boldsymbol{r}') = 0 . \tag{16.3}$$

General semi-microscopic boundary conditions on surfaces and interfaces cannot be satisfied by the ansatz (14.1) in which higher harmonics with respect to the velocity vector were neglected. It has, however, to be expected that scattering processes will produce an almost isotropic velocity distribution already close to surfaces and interfaces. The diffusion equation, which was derived on the basis of Eq. (14.1), should, therefore, be valid everywhere in the sample with the exception of surface and interface layers with a thickness of the order of the electronic mean free path (or, rather, the transport length). The problem of formulating boundary conditions for solutions to the diffusion equation is again a problem of extrapolating the solutions as such through these surface and interface layers. This extrapolation procedure shall be performed below with the help of an appropriate variational principle [13]. Eqs. (16.2) and (16.3) will be confirmed on external surfaces for arbitrary reflexion kernels. Here, we only want to emphasize that the diffusion approximation can only be valid if all linear dimensions (including radii of curvature) of the sample are big as compared to the electronic mean free path.

A GL like formulation of the diffusion approximation can also be given for a finite sample provided that it consists of a single metal. The considerations of Section 14c, in particular Eq. (14.30), are still valid in this case if the eigenfunctions $\varphi_m(r)$ are defined by the differential equation (14.25) and the boundary condition

$$n \cdot \tilde{\partial} \varphi_m(r) = 0 . \tag{16.4}$$

This argument leads to the GL equation (13.1) or (13.35) and the boundary condition (15.1). No such formulation is possible if the sample consists of several metals. Eq. (15.1) is, however, still true on external surfaces, but now as a consequence of the gap equation (10.1) or (10.2) and the boundary condition (16.3).

The general considerations given in small print immediately before Section 15a may also be made here in a similar way.

### a) Variational Principle

Eq. (11.23) shall first be reformulated with the help of an appropriate Green's function, $e_\omega^{(tr)}(r, v; r', v')$. It is defined as the (unique) solution to the equation

$$(2|\omega| + v \cdot \tilde{\partial} + nv\sigma_{tr}) \, e_\omega^{(tr)}(r, v; r', v') = \delta(r - r') \, \hat{\delta}(v, v') \tag{16.5}$$

and the correct semi-microscopic boundary conditions (in terms of the variables $r$ and $v$) on surfaces and interfaces. The transport cross section, $\sigma_{tr}$, is given by Eq. (13.22). Eq. (11.23) plus boundary conditions may now be replaced by the integral equation

$$\Delta_\omega(r, v) = \Delta_\omega^{(tr)}(r, v) + \int d^3r' \, d\Omega' \, d\Omega'' \, e_\omega^{(tr)}(r, v; r', v') \, w(r'; v', v'') \Delta_\omega(r', v'') \tag{16.6}$$

with

$$\Delta_\omega^{(tr)}(r, v) = \frac{1}{(2\pi)^2} \int d^3r' \, d\Omega' \, e_\omega^{(tr)}(r, v; r', v') \, \Delta(r') \tag{16.7}$$

and

$$w(r'; v', v'') = nv \left[ \frac{d\sigma(\theta)}{d\Omega} - \sigma^{(1)} \hat{\delta}(v', v'') \right] = w(r'; -v'', -v') . \tag{16.8}$$

$\theta$ is the angle between the velocity vectors $v''$ and $v'$. The scattering cross section $\sigma^{(1)}$ is given by Eq. (13.20). It is

$$\oint d\Omega' w(r'; v', v'') = \oint d\Omega'' w(r'; v', v'') = nv\sigma_{tr} \tag{16.9}$$

and

$$\oint d\Omega' v' w(r'; v', v'') = \oint d\Omega'' v'' w(r'; v', v'') = 0 . \tag{16.10}$$

Eq. (16.10) will turn out to be particularly useful.

The Green's function satisfies the relation

$$N(r)\, e_\omega^{(\mathrm{tr})}(r, v; r', v') = N(r')\, e_\omega^{(\mathrm{tr})*}(r', -v'; r, -v)\,. \tag{16.11}$$

Here, $N(r)$ is the density of states of that metal in which the point $r$ is situated. For the proof of Eq. (16.11), Eq. (16.5) is multiplied by $N(r)\, e_\omega^{(\mathrm{tr})*}(r, -v; r'', -v'')$. Then the complex conjugate of the equation for $e_\omega^{(\mathrm{tr})}(r, -v; r'', -v'')$ is multiplied by $N(r)\, e_\omega^{(\mathrm{tr})}(r, v; r', v')$. The difference of both equations is integrated with respect to $v$ over all directions and with respect to $r$ over the whole sample. Only boundary contributions remain on the lefthand side which, however, vanish as a consequence of Eqs. (10.16), (10.19), (10.26), (10.29), and (10.30). The righthand side leads to Eq. (16.11) apart from some notational differences.

An integral equation for the function $\oint w(r; v, v')\, \Delta_\omega(r, v')\, \mathrm{d}\Omega'$ is easily derived from Eq. (16.6). It reads

$$\oint \mathrm{d}\Omega'\, w(r; v, v')\, \Delta_\omega(r, v') = \oint \mathrm{d}\Omega'\, w(r; v, v')\, \Delta_\omega^{(\mathrm{tr})}(r, v') \tag{16.12}$$
$$+ \int \mathrm{d}^3 r'\, \mathrm{d}\Omega'\, \mathrm{d}\Omega''\, w(r; v, v')\, e_\omega^{(\mathrm{tr})}(r, v'; r', v'')\, w(r'; v'', v''')\, \Delta_\omega(r', v''')\,.$$

This integral equation is the stationarity condition of the functional

$$4\pi I_\omega = \int \left[\Delta_\omega^*(r, -v)\, \Delta_\omega(r, v') - \Delta_\omega^*(r, -v)\, \Delta_\omega^{(\mathrm{tr})}(r, v') - \Delta_\omega^{(\mathrm{tr})*}(r, -v)\, \Delta_\omega(r, v')\right] \cdot$$
$$\cdot w(r; v, v')\, N(r)\, \mathrm{d}^3 r\, \mathrm{d}\Omega\, \mathrm{d}\Omega' - \int \Delta_\omega^*(r, -v)\, w(r; v, v')\, N(r) \tag{16.13}$$
$$e_\omega^{(\mathrm{tr})}(r, v'; r', v'')\, w(r'; v'', v''')\, \Delta_\omega(r', v''')\, \mathrm{d}^3 r\, \mathrm{d}^3 r'\, \mathrm{d}\Omega\, \mathrm{d}\Omega'\, \mathrm{d}\Omega''\, \mathrm{d}\Omega'''\,.$$

Here, $\Delta_\omega^{(\mathrm{tr})}(r, v)$ is assumed to be a given function. $I_\omega$ is a real functional because of Eqs. (16.11) and (16.8). Varying only $\Delta_\omega^*(r, -v)$ we are immediately lead to Eq. (16.12).

The integral equation (16.12) and the variational principle based upon the functional (16.13) are particularly suited to the diffusion approximation. With the ansatz (14.1), it is

$$\int w(r; v, v')\, \Delta_\omega(r, v')\, \mathrm{d}\Omega' = n v \sigma_{\mathrm{tr}} H_\omega(r)\,, \tag{16.14}$$

where use was made of Eqs. (16.9) and (16.10). Eq. (16.5) may formally be solved in an iterative manner with the result

$$e_\omega^{(\mathrm{tr})}(r, v; r', v') = \frac{1}{2|\omega| + n v \sigma_{\mathrm{tr}}} \sum_{l=0}^{\infty} \left(-v \cdot \tilde{\partial}\, \frac{1}{2|\omega| + n v \sigma_{\mathrm{tr}}}\right)^l \delta(r - r')\, \hat{\delta}(v, v')\,. \tag{16.15}$$

A possible position dependence of the impurity concentration, $n$, shall from now on be disregarded though it could be kept at all steps of the calculation. Keeping only the $l = 0$ term of Eq. (16.15) and making use of Eq. (16.9) we obtain

$$\oint \mathrm{d}\Omega'\, w(r; v, v')\, \Delta_\omega^{(\mathrm{tr})}(r, v') = \frac{n v \sigma_{\mathrm{tr}}}{(2\pi)^2\, (2|\omega| + n v \sigma_{\mathrm{tr}})}\, \Delta(r) + \cdots \tag{16.16}$$

from Eq. (16.7).

Both the lefthand side and the first term on the righthand side of Eq. (16.12) in this way become independent of $v$. The second term on the righthand side becomes velocity independent only after an average with respect to the directions of $v$ has been performed. If we do not go beyond the second derivative in Eq. (16.15), we obtain

$$\frac{1}{4\pi} \int d^3 r' d\Omega d\Omega' d\Omega'' d\Omega''' w(r; v, v') e_\omega^{(tr)}(r, v'; r', v'') w(r'; v'', v''') \Delta_\omega(r', v''')$$

$$= \frac{(nv\sigma_{tr})^2}{2|\omega| + nv\sigma_{tr}} \left(1 + \frac{v^2}{3(2|\omega| + nv\sigma_{tr})^2} \tilde{\partial} \cdot \tilde{\partial} + \cdots \right) H_\omega(r). \qquad (16.17)$$

Now everything is inserted into Eq. (16.12). After the non-derivative terms have been combined, $2|\omega| + nv\sigma_{tr}$ may be replaced by $nv\sigma_{tr}$ (dirty limit!). In this way, the diffusion equation (14.8) is obtained.

The same steps may be made in the functional (16.13). Making use of Eqs. (16.14) and (16.16) we derive

$$I_\omega = \int \left[ nv\sigma_{tr} |H_\omega(r)|^2 - \frac{1}{(2\pi)^2} (H_\omega^*(r) \Delta(r) + \Delta^*(r) H_\omega(r)) \right] N(r) d^3 r \qquad (16.18)$$

$$- \frac{1}{4\pi} \int nv\sigma_{tr} H_\omega^*(r) N(r) e_\omega^{(tr)}(r, v; r'v') nv\sigma_{tr} H_\omega(r') d^3 r d^3 r' d\Omega d\Omega' ,$$

which is a real functional of $H_\omega(r)$ for given $\Delta(r)$. With the help of the definition

$$E_\omega(r, v) = \int e_\omega^{(tr)}(r, v; r', v') nv\sigma_{tr} H_\omega(r') d^3 r' d\Omega' , \qquad (16.19)$$

Eq. (16.18) may be rewritten in the form

$$I_\omega = \int \left[ nv\sigma_{tr} |H_\omega(r)|^2 - \frac{1}{(2\pi)^2} (H_\omega^*(r) \Delta(r) + \Delta^*(r) H_\omega(r)) \right] N(r) d^3 r \qquad (16.20)$$

$$- \frac{1}{4\pi} \int nv\sigma_{tr} H_\omega^*(r) N(r) E_\omega(r, v) d^3 r d\Omega .$$

The further analysis proceeds in essentially the same way as in Section 15 in connection with the functional (15.5). We put

$$E_\omega(r, v) = E_{\omega 0}(r, v) + E_{\omega 1}(r, v) , \qquad (16.21)$$

where $E_{\omega 0}(r, v)$ is calculated with the help of the righthand side of Eq. (16.15),

$$E_{\omega 0}(r, v) = \left[ \left(1 - \frac{2|\omega|}{nv\sigma_{tr}} \right) - \frac{1}{nv\sigma_{tr}} v \cdot \tilde{\partial} + \frac{1}{(nv\sigma_{tr})^2} (v \cdot \tilde{\partial})^2 + \cdots \right] H_\omega(r) .$$

$$(16.22)$$

Here, simplifications corresponding to the dirty limit have already been introduced. Both $E_\omega(r, v)$ and $E_{\omega 0}(r, v)$ satisfy the differential equation

$$(2|\omega| + v \cdot \tilde{\partial} + nv\sigma_{\mathrm{tr}}) E_\omega(r, v) = nv\sigma_{\mathrm{tr}} H_\omega(r),\qquad(16.23)$$

which follows from Eq. (16.5), so that $E_{\omega 1}(r, v)$ obeys the corresponding homogeneous differential equation. $E_{\omega 1}(r, v)$ satisfies inhomogeneous boundary conditions on surfaces and interfaces in order that the right-hand side of Eq. (16.12) satisfies the correct semi-microscopic boundary conditions. $E_{\omega 1}(r, v)$ is, in the dirty limit, markedly different from zero only in a boundary layer of approximate thickness $l_{\mathrm{tr}}$.

Eq. (16.20) is now rewritten in the form

$$I_\omega = I_{\omega 0} + I_{\omega 1},\qquad(16.24)$$

which corresponds to Eq. (15.15). In Eq. (16.24), it is

$$I_{\omega 0} = \int \left\{ H_\omega^*(r) \cdot \left[ 2|\omega| H_\omega(r) - \frac{vl_{\mathrm{tr}}}{3} \tilde{\partial} \cdot \tilde{\partial} H_\omega(r) \right] \right. \qquad(16.25)$$
$$\left. - \frac{1}{(2\pi)^2} (H_\omega^*(r) \Delta(r) + \Delta^*(r) H_\omega(r)) \right\} N(r)\, d^3 r$$

and

$$I_{\omega 1} = -\frac{1}{4\pi} \int nv\sigma_{\mathrm{tr}} H_\omega^*(r) N(r) E_{\omega 1}(r, v)\, d^3 r\, d\Omega.\qquad(16.26)$$

Variation of $I_{\omega 0}$ with respect to $H_\omega^*(r)$ leads to the diffusion equation (14.8). The integrand of $I_{\omega 1}$ is different from zero only in a boundary layer of approximate thickness $l_{\mathrm{tr}}$. $I_{\omega 1}$ may be transformed into a boundary integral proper,

$$I_{\omega 1} = \frac{1}{4\pi} \sum_j N_j \oint d\Omega \oint_{B_j} d^2 r \cdot v_j\, E_{\omega 0}^*(r, -v_j) E_{\omega 1}(r, v_j),\qquad(16.27)$$

in the same way as this was done in Eqs. (15.17) and (15.59). Boundary conditions will be obtained from varying $I_{\omega 1}$ with respect to $H_\omega^*(r)$.

The ansatz (14.1) is in general not valid in the boundary layers. But in the spirit of the extrapolation procedure it is used everywhere in the sample.

### b) Boundary Conditions

On a surface of the sample we may put

$$E_{\omega 1}(r, v_{\mathrm{in}}) = 0\qquad(16.28)$$

in complete analogy to Eq. (15.19) provided that all linear dimensions of the sample (and even of each metal in the sample) are big as compared to the transport length. The inhomogeneous boundary condition to be satisfied by $E_{\omega 1}(r, v)$ leads to

$$E_{\omega 1}(r, v_{\text{out}}) = -E_{\omega 0}(r, v_{\text{out}}) + \int R(r; v_{\text{out}}, v_{\text{in}})\, E_{\omega 0}(r, v_{\text{in}})\, d\Omega_{\text{in}} \quad (16.29)$$

which corresponds to Eq. (15.20). If an expansion analogous to Eq. (15.23),

$$I_{\omega 1} = \sum_{l, m=0}^{\infty} I_{\omega 1}(l, m), \quad (16.30)$$

is made, the only non-vanishing surface contribution up to $l + m = 1$ is given by

$$I_{\omega 1}(0, 1) = N \int \frac{v l_{\text{tr}}}{3} H_{\omega}^{*}(r)\, d^2 r \cdot \tilde{\partial} H_{\omega}(r). \quad (16.31)$$

Variation with respect to $H_{\omega}^{*}(r)$ confirms the boundary condition (16.2).

On an interface between metal 1 and metal 2, we may put

$$E_{\omega 1}(r, v_{1\,\text{in}}) = 0 = E_{\omega 1}(r, v_{2\,\text{in}}). \quad (16.32)$$

The inhomogeneous boundary conditions to be satisfied by $E_{\omega 1}(r, v)$ lead to

$$\begin{aligned}
E_{\omega 1}(r, v_{1\,\text{out}}) = &-E_{\omega 0}(r, v_{1\,\text{out}}) + \int R_1(r; v_{1\,\text{out}}, v_{1\,\text{in}})\, E_{\omega 0}(r, v_{1\,\text{in}})\, d\Omega_{1\,\text{in}} \\
&+ \int T_1(r; v_{1\,\text{out}}, v_{2\,\text{in}})\, E_{\omega 0}(r, v_{2\,\text{in}})\, d\Omega_{2\,\text{in}}
\end{aligned} \quad (16.33)$$

and to an analogous equation with subscripts 1 and 2 interchanged. These equations correspond to Eqs. (15.60) and (15.61), respectively. All further steps are made in exactly the same manner as in Section 15c. We find

$$\left\{ \begin{aligned}
I_{\omega 1}(0, 0) &= \frac{1}{4} \int |d^2 r|\, N_1\, v_1\, t_1\, |H_{1\omega}(r) - H_{2\omega}(r)|^2, \\
I_{\omega 1}(1, 0) &= -\frac{1}{2} \int d^2 r \cdot \left( N_1 \frac{v_1 l_{1\text{tr}}}{3} t_1' \tilde{\partial} H_{1\omega}(r) + N_2 \frac{v_2 l_{2\text{tr}}}{3} t_2' \tilde{\partial} H_{2\omega}(r) \right)^{*} \\
&\qquad\qquad\qquad\qquad\qquad\qquad\qquad \cdot (H_{1\omega}(r) - H_{2\omega}(r)), \\
I_{\omega 1}(0, 1) &= I_{\omega 1}^{*}(1, 0) + \int d^2 r \\
&\qquad \cdot \left( N_1 \frac{v_1 l_{1\text{tr}}}{3} H_{1\omega}^{*}(r) \tilde{\partial} H_{1\omega}(r) - N_2 \frac{v_2 l_{2\text{tr}}}{3} H_{2\omega}^{*}(r) \tilde{\partial} H_{2\omega}(r) \right).
\end{aligned} \right. \quad (16.34)$$

$H_{1\omega}(r)$ and $H_{2\omega}(r)$ are the boundary values of the isotropic parts of the distribution functions in metal 1 and 2, respectively. The coefficients $t_1$, $t_1'$, and $t_2'$ to some extent characterize the transmission properties of the interface in the point $r$. They were defined in Eqs. (15.73) and (15.77).

They satisfy certain inequalities which were given in Section 15c. Further it is

$$N_1 v_1 t_1 = N_2 v_2 t_2 ; \tag{16.35}$$

cf. Eq. (15.64).

The complete functional given by Eq. (16.24) is still real. It is, therefore, permitted to vary $H_{j\omega}(r)$ and $H_{j\omega}^*(r)$ independently. Varying $H_{1\omega}^*(r)$ and $H_{2\omega}^*(r)$ in the sum of the contributions to $I_{\omega 1}$ written down in Eq. (16.34), we obtain the boundary conditions

$$\frac{N_1 v_1 t_1}{2(2 - t_1' - t_2')} (H_{2\omega}(r) - H_{1\omega}(r)) = N_1 \frac{v_1 l_{1\mathrm{tr}}}{3} \, \boldsymbol{n} \cdot \tilde{\partial} H_{1\omega}(r)$$
$$= N_2 \frac{v_2 l_{2\mathrm{tr}}}{3} \, \boldsymbol{n} \cdot \tilde{\partial} H_{2\omega}(r) \tag{16.36}$$

on an interface between metal 1 and metal 2. $\boldsymbol{n}$ is the interface normal unit vector pointing into metal 2. It is

$$\frac{N_1 v_1 t_1}{2 - t_1' - t_2'} \geqq 0 ; \tag{16.37}$$

cf. Eq. (15.89).

Contributions from the variation of $[\boldsymbol{n} \cdot \tilde{\partial} H_{j\omega}(r)]^*$ may again be neglected for reasons given in Section 15c. If the denominator in Eq. (16.36) vanishes, we obtain the boundary conditions

$$H_{1\omega}(r) = H_{2\omega}(r) ,$$
$$N_1 v_1 l_{1\mathrm{tr}} \boldsymbol{n} \cdot \tilde{\partial} H_{1\omega}(r) = N_2 v_2 l_{2\mathrm{tr}} \boldsymbol{n} \cdot \tilde{\partial} H_{2\omega}(r) . \tag{16.38}$$

Vanishing of the denominator is strictly possible only if Eq. (6.25) holds. The boundary conditions (16.38) shall, however, also be used in other cases. Then they will be referred to as simplified boundary conditions. It is hoped that the simplified boundary conditions will lead to reasonable results provided that the electric contact through the interface is sufficiently good.

If a sample consists of several metals in contact it does not seem possible to give a GL like formulation for the determination of the pair potential. Eq. (14.9) has still to be solved. The cut-off, symbolized by the dash, has to be done at the Debye frequency of that metal which is present at the point $r$. The symmetric cut-off of Eq. (10.2) is introduced by putting $\Delta(r)$ equal to zero in Eq. (14.8) if $|\omega|$ is bigger than the Debye frequency at the point $r$.

The boundary conditions for $H_\omega(r)$ may be translated into boundary conditions for $K_\omega(r, r')$ which have to be used in conjunction with

10*

Eq. (14.10). It follows from Eq. (16.36) that it is

$$
\frac{N_1 v_1 t_1}{2(2 - t_1' - t_2')} \left( \frac{K_{2\omega}(r, r')}{N_2} - \frac{K_{1\omega}(r, r')}{N_1} \right)
$$

$$
= \frac{v_1 l_{1tr}}{3} \, n \cdot \tilde{\partial} K_{1\omega}(r, r') = \frac{v_2 l_{2tr}}{3} \, n \cdot \tilde{\partial} K_{2\omega}(r, r') .
$$

$(16.39)$

The simplified boundary conditions (16.38) go over into

$$
\frac{K_{1\omega}(r, r')}{N_1} = \frac{K_{2\omega}(r, r')}{N_2} ,
$$

$(16.40)$

$$
v_1 l_{1tr} n \cdot \tilde{\partial} K_{1\omega}(r, r') = v_2 l_{2tr} n \cdot \tilde{\partial} K_{2\omega}(r, r') .
$$

These simplified boundary conditions (16.40) were proposed earlier by *de Gennes* [1].

   The proofs of the uniqueness theorem and of general properties of the kernel, as given in Section 14b for the diffusion approximation, are now easily completed. The contributions from surfaces to the lefthand side of Eq. (14.18) vanish because of Eq. (16.3). On interfaces, the contributions from the two adjoining metals are combined. These contributions are never positive as a consequence of Eqs. (16.39) and (16.37). They vanish for the simplified boundary conditions (16.40). So the proof of the uniqueness theorem is made complete. That all contributions to Eq. (14.19) vanish is also easily seen. Finally, also Eq. (14.24) is shown to be correct.

For practical calculations it is again useful to have a variational principle at hand which does not suffer from derivative terms in the interface contributions. Such a variational principle corresponds to the one derived in Section 15d. The appropriate functional is now given by

$$
I_\omega' = \sum_j N_j \int_{V_j} \cdot \left[ 2|\omega| \, |H_{j\omega}(r)|^2 + \frac{v_j l_{jtr}}{3} \, |\tilde{\partial} H_{j\omega}(r)|^2 \right.
$$

$$
\left. - \frac{1}{(2\pi)^2} \, (H_{j\omega}^*(r) \, \Delta(r) + \Delta^*(r) \, H_{j\omega}(r)) \right] \cdot d^3 r
$$

$(16.41)$

$$
+ \int B_{12}(r) \, |H_{2\omega}(r) - H_{1\omega}(r)|^2 \, |d^2 r| ,
$$

where it is

$$
B_{12}(r) := \frac{N_1 v_1 t_1(r)}{2(2 - t_1'(r) - t_2'(r))} \geq 0 .
$$

$(16.42)$

If the simplified boundary conditions (16.38) are assumed, the boundary integral in Eq. (16.41) has to be discarded. Instead,

$$
H_{1\omega}(r) = H_{2\omega}(r)
$$

$(16.43)$

is postulated in any point on the interface.

## 17. Single Homogeneous Metal with Isotropic Impurity Scattering

### a) General Formulations

A special approximation scheme can be formulated [14, 45] for a sample consisting of a single homogeneous [i.e., $n(r) = $ const.] metal with isotropic impurity scattering,

$$\frac{d\sigma(\theta)}{d\Omega} = \frac{\sigma}{4\pi}.$$ (17.1)

This approximation contains the GL approximation (Sections 13 and 15) and the diffusion approximation (Sections 14 and 16) as special cases. It further permits to go beyond the standard boundary condition (15.1) even in the presence of impurities. The assumption of isotropic impurity scattering is presumably not very realistic. It is, nevertheless, hoped that at least the general trend of the influence of impurity scattering will be represented correctly, even if the basic assumption is not correct. The situation is somewhat similar to that in the treatment of Fermi surfaces: instead of working with the correct Fermi surface, we derive results under the assumption of a spherical Fermi surface, and apply the results to real metals.

Before introducing approximations, we shall derive results which are strictly valid under the assumptions that the impurity concentration (or the electronic mean free path) is position independent and that the scattering cross section is isotropic. We mention that it is $\sigma_{tr} = \sigma$ and $l_{tr} = l$ in this case. Eq. (12.16) reduces to

$$g_\omega(r, v; r') = g_\omega^{(\sigma)}(r, v; r') \\ + n v \sigma \pi \int d^3 r'' \, d\Omega' \, g_\omega^{(\sigma)}(r, v; r'') \, g_\omega(r'', v'; r'),$$ (17.2)

where the distribution function $g_\omega^{(\sigma)}(r, v; r')$ has been defined by Eq. (12.11), with the semi-microscopic boundary condition (10.16) satisfied on the surface of the sample. There are no interfaces under the assumption that the sample consists of a single metal. The distribution function, $g_\omega^{(\sigma)}(r, v; r')$, represents those electrons (or carriers of information), which travel from $r'$ to $r$ with velocity $v$ without suffering a single impurity collision.

The $v'$ integration may be done in Eq. (17.2). If this equation is further integrated with respect to $v$ over all directions, we obtain [46]

$$K_\omega(r, r') = K_\omega^{(\sigma)}(r, r') + \frac{v}{2\pi N l} \int K_\omega^{(\sigma)}(r, r'') \, K_\omega(r'', r') \, d^3 r'', \quad (17.3)$$

where the kernel $K_\omega(r, r')$ was introduced with the help of Eq. (10.4). Analogously, $K_\omega^{(\sigma)}(r, r')$ is defined by

$$K_\omega^{(\sigma)}(r, r') = 2\pi^2 N \oint g_\omega^{(\sigma)}(r, v; r') \, d\Omega.$$ (17.4)

The electronic mean free path, $l$, was taken from Eq. (12.15). Introducing operators according to Eq. (11.17), we may rewrite Eq. (17.3) in the form

$$\mathbb{K}_\omega = \mathbb{K}_\omega^{(\sigma)} + \frac{v}{2\pi N l}\, \mathbb{K}_\omega^{(\sigma)}\, \mathbb{K}_\omega\,. \qquad (17.5)$$

Eq. (17.5) may formally be solved for $\mathbb{K}_\omega$ with the result

$$\mathbb{K}_\omega = \frac{\mathbb{K}_\omega^{(\sigma)}}{1 - \dfrac{v}{2\pi N l}\, \mathbb{K}_\omega^{(\sigma)}}\,. \qquad (17.6)$$

The Hermitian operators $\mathbb{K}_\omega$ and $\mathbb{K}_\omega^{(\sigma)}$ have a common complete set of eigenfunctions. If it is

$$\mathbb{K}_\omega^{(\sigma)} \Delta(r) = \frac{2\pi N}{v}\, \kappa_\omega^{(\sigma)} \Delta(r)\,, \qquad (17.7)$$

it also is

$$\mathbb{K}_\omega \Delta(r) = \frac{2\pi N}{v}\, \kappa_\omega \Delta(r) \qquad (17.8)$$

with the eigenvalue $\kappa_\omega$ given by

$$\kappa_\omega = \frac{\kappa_\omega^{(\sigma)}}{1 - \dfrac{\kappa_\omega^{(\sigma)}}{l}}\,. \qquad (17.9)$$

The denominators in Eqs. (17.6) and (17.9) are not dangerous. A straightforward generalization of the arguments given in Section 11b shows that it is

$$0 < \mathbb{K}_\omega^{(\sigma)} \leqq \frac{2\pi N}{2|\omega| + n v \sigma} < \frac{2\pi N l}{v}\,; \qquad (17.10)$$

cf. Eq. (11.19). It follows from this equation that it also is

$$0 < \kappa_\omega^{(\sigma)} < l\,. \qquad (17.11)$$

## b) Approximations

Now we turn to an approximative evaluation of the kernel $K_\omega^{(\sigma)}(r, r')$ and to an approximative determination of eigenfunctions and eigenvalues. The distribution function,

$$\Delta_\omega^{(\sigma)}(r, v) = \int g_\omega^{(\sigma)}(r, v; r')\, \Delta(r')\, d^3 r'\,, \qquad (17.12)$$

satisfies the differential equation

$$(2|\omega| + v \cdot \tilde{\partial} + n v \sigma)\, \Delta_\omega^{(\sigma)}(r, v) = \frac{1}{(2\pi)^2}\, \Delta(r) \qquad (17.13)$$

and the correct semi-microscopic boundary conditions. Eq. (17.13) is solved in an iterative manner which is known from Section 13a. The result is

$$\Delta_\omega^{(\sigma)}(\mathbf{r}, \mathbf{v}) = \frac{1}{(2\pi)^2} \sum_{p=0}^{\infty} \frac{(-\mathbf{v} \cdot \tilde{\partial})^p}{(2|\omega| + n v \sigma)^{p+1}} \Delta(\mathbf{r}) \, ; \qquad (17.14)$$

cf. Eq. (13.4). The expansion is justified and may be cut off at the appropriate place, if $\Delta(\mathbf{r})$ is a slowly varying function on the range of the kernel $K_\omega^{(\sigma)}(\mathbf{r}, \mathbf{r}')$. The latter is estimated by $\zeta_\omega$ [Eq. (12.14)]. If we do not go beyond the second derivative in Eq. (17.14), we obtain

$$\int K_\omega^{(\sigma)}(\mathbf{r}, \mathbf{r}') \, \Delta(\mathbf{r}') \, \mathrm{d}^3 r' = \frac{2\pi N \zeta_\omega}{v} \left( 1 + \frac{\zeta_\omega^2}{3} \tilde{\partial} \cdot \tilde{\partial} \right) \Delta(\mathbf{r}) \qquad (17.15)$$

with the help of essentially Eq. (11.24) and of Eq. (13.6). The length $\zeta_\omega$ was introduced from Eq. (12.14).

In an infinite metal where no boundary conditions apart from the boundedness have to be satisfied, it follows from Eq. (17.15) that $\Delta(\mathbf{r})$ is an eigenfunction to $\mathbb{K}_\omega^{(\sigma)}$, if it satisfies a GL like differential equation,

$$\tilde{\partial} \cdot \tilde{\partial} \, \Delta(\mathbf{r}) = -\lambda \Delta(\mathbf{r}) \, . \qquad (17.16)$$

The eigenvalue, $\kappa_\omega^{(\sigma)}$, is then given by

$$\kappa_\omega^{(\sigma)} = \zeta_\omega \left( 1 - \frac{\lambda \zeta_\omega^2}{3} \right) . \qquad (17.17)$$

The condition, that $\Delta(\mathbf{r})$ is a sufficiently slowly varying function, may be written as

$$\lambda \zeta_\omega^2 \ll 1 \, . \qquad (17.18)$$

This condition is, at a given temperature, satisfied for all values of $\omega$ appearing in the gap equation (10.1), if it is

$$\lambda \zeta_0^2(T) \ll 1 \qquad (17.19)$$

with $\zeta_0(T)$ given by

$$\frac{1}{\zeta_0(T)} := \frac{1}{\xi_0(T)} + \frac{1}{l} \, . \qquad (17.20)$$

$\xi_0(T)$ has been defined in Eq. (13.11).

In a finite sample, boundary conditions have to be satisfied on the surface. The boundary condition is derived by remembering, that $(\Delta, \mathbb{K}_\omega^{(\sigma)} \Delta)$ is stationary with respect to variations of $\Delta(\mathbf{r})$ with $(\Delta, \Delta)$ kept constant, if and only if $\Delta(\mathbf{r})$ is an eigenfunction to $\mathbb{K}_\omega^{(\sigma)}$. In

$$(\Delta, \mathbb{K}_\omega^{(\sigma)} \Delta) = 2\pi^2 N \int \Delta^*(\mathbf{r}) \, \Delta_\omega^{(\sigma)}(\mathbf{r}, \mathbf{v}) \, \mathrm{d}^3 r \, \mathrm{d}\Omega \, , \qquad (17.21)$$

the same procedures may be applied as in Section 15. The extrapolation has now to be made through a surface layer of approximate thickness $\zeta_\omega$. We put

$$\Delta_\omega^{(\sigma)}(r, v) = \Delta_{\omega 0}^{(\sigma)}(r, v) + \Delta_{\omega 1}^{(\sigma)}(r, v) \tag{17.22}$$

with $\Delta_{\omega 0}^{(\sigma)}(r, v)$ given by the righthand side of Eq. (17.14). $\Delta_{\omega 1}^{(\sigma)}(r, v)$ obeys Eq. (17.13) with the righthand side put equal to zero, and inhomogeneous boundary conditions. Exactly the same steps as made in Section 15a lead to

$$(\Delta, \mathbb{K}_\omega^{(\sigma)}\Delta) = \frac{2\pi N \zeta_\omega}{v} \int \left[ |\Delta(r)|^2 - \frac{\zeta_\omega^2}{3} |\tilde{\partial} \Delta(r)|^2 \right] d^3 r \; ; \tag{17.23}$$

cf. Eq. (15.14). The righthand side is real as it must for an expectation value of a Hermitian operator. Varying $\Delta^*(r)$ for constant $(\Delta, \Delta)$, we obtain the Eqs. (17.16) and (17.17), and the boundary condition (15.1).

Differential equation (17.16) and boundary condition (15.1) are independent of $\omega$. A function $\Delta(r)$ satisfying this differential equation and boundary condition, therefore, is a simultaneous eigenfunction to all kernels $\mathbb{K}_\omega^{(\sigma)}$ and so to all kernels $\mathbb{K}_\omega$. It is

$$\kappa_\omega = \frac{1 - \dfrac{\lambda \zeta_\omega^2}{3}}{\dfrac{1}{\zeta_\omega} + \dfrac{\lambda \zeta_\omega^2}{3l}} = v \frac{1 - \dfrac{\lambda \zeta_\omega^2}{3}}{2|\omega| + \dfrac{\lambda \zeta_\omega^2 v}{3l}} . \tag{17.24}$$

Eq. (17.9), (17.17), (12.14), and (12.4) were used in the course of the derivation.

The gap equation (10.1) is simplified if the pair potential is an eigenfunction to each individual kernel, $\mathbb{K}_\omega$. It reads

$$1 = \frac{2\pi g N}{v} T \sum_\omega{}' \kappa_\omega . \tag{17.25}$$

Inserting Eq. (17.24) and making use of Eq. (13.9), we obtain

$$\ln\left(\frac{T_c}{T}\right) = 2\pi T \sum_\omega \left( \frac{1}{2|\omega|} - \frac{1 - \dfrac{\lambda \zeta_\omega^2}{3}}{2|\omega| + \dfrac{\lambda \zeta_\omega^2 v}{3l}} \right) \tag{17.26}$$

where it has still to be checked whether the cut-off may be disregarded. This equation may, because of Eq. (17.18), also be written as

$$\ln\left(\frac{T_c}{T}\right) = 2\pi T \sum_\omega \left( \frac{1}{2|\omega|} - \frac{1}{2|\omega| + \dfrac{\lambda \zeta_\omega v}{3}} \right) \tag{17.27}$$

That Eqs. (17.26) and (17.27) define $\lambda$ uniquely as a non-negative function of temperature for $T \leq T_c$ is seen in the same way as in Section 14c.

Diffusion approximation and GL approximation are contained as special cases in Eq. (17.27). This is immediately clear for the diffusion approximation since it is $\zeta_\omega \approx l$ for $l \ll \xi_\omega$; in this way Eq. (14.31) is obtained. That $l \ll \xi_\omega$ is not true for arbitrarily high values of $|\omega|$ was already pointed out and discussed regarding its consequences at the end of Section 14c. The GL approximation is obtained for sufficiently small values of $\lambda$, more exactly for

$$\lambda \xi_\omega \zeta_\omega \ll 3 . \tag{17.28}$$

If a series expansion is made in the second term on the righthand side of Eq. (17.27), we derive

$$\ln\left(\frac{T_c}{T}\right) = \frac{\lambda v^2}{3} 2\pi T \sum_\omega \frac{1}{(2|\omega|)^2 (2|\omega| + n v \sigma)} . \tag{17.29}$$

This equation is identical with the GL result [Eq. (13.26)] if the logarithm is replaced by the first term of a series expansion for $T \approx T_c$ and if $\sigma_{tr} = \sigma$ is remembered. Incidentally, the length $\sqrt{\xi_\omega \zeta_\omega}$ occurring in Eq. (17.28) is a reasonable interpolation between the ranges of $K_\omega(r, r')$ in the clean ($\xi_\omega$) and the dirty ($\sqrt{\xi_\omega l}$) limit.

We did not study the problem how far the region of validity of the present approximation extends beyond those of the GL approximation ($T \approx T_c$) and of the diffusion approximation ($l \ll \xi_0(T_c)$).

### c) Perturbation Treatment

Now we want to go beyond the approximation of Section 17b, which lead to the GL equation (13.1) and the boundary condition (15.1). The quantity $\lambda(T)$ was defined with the help of Eq. (17.26) or the approximation to it, Eq. (17.27). For $T \approx T_c$, $\lambda(T)$ may be taken from the GL approximation (with $\sigma_{tr}$ replaced by $\sigma$ because of isotropic scattering). In the dirty limit [Eq. (14.15)], it may be obtained from the diffusion approximation.

We apply a perturbation treatment based upon the functional

$$\Phi = \frac{1}{g} (\Delta, \Delta) - T \sum_\omega{}' (\Delta, \mathbb{K}_\omega \Delta) , \tag{17.30}$$

which was given in Eq. (2.20). $\Phi$ is equal to zero and stationary with respect to variations of $\Delta(r)$ at a second kind transition. It, therefore, is stationary if Eqs. (17.8) and (17.25) hold. If we go beyond the approximation given in Section 17b, the approximate expression for the kernel is altered by a small amount,

$$\mathbb{K}_\omega \to \mathbb{K}_\omega + \delta \mathbb{K}_\omega . \tag{17.31}$$

In the lowest order perturbative approximation, we have

$$T \sum_\omega (\Delta, \delta \mathbb{K}_\omega \Delta) = 0 . \tag{17.32}$$

The inevitable change of $\Delta(r)$ does not contribute to the result because of the stationarity of the functional (17.30).

If $\lambda(T)$ is not calculated from Eq. (17.26), but from approximations to it (GL approximation, diffusion approximation), Eq. (17.32) is replaced by

$$T \sum (\Delta, \delta \, \mathbb{K}_\omega \Delta) + C(T)(\Delta, \Delta) = 0 \, . \tag{17.33}$$

$C(T)$ is a correction term due to the additional approximation made in the calculation of $\lambda(T)$. We shall, however, try to make only those applications in Chapter III, in which $C(T)$ may be neglected. There are in principle further unknown corrections if the differential scattering cross section is not isotropic (and the Fermi surface not spherical).

The operator $\delta \, \mathbb{K}_\omega$ entering into Eq. (17.32) may be expressed by the operator $\delta \, \mathbb{K}_\omega^{(\sigma)}$. Indeed, it follows from Eq. (17.5) that it is

$$\delta \, \mathbb{K}_\omega = \delta \, \mathbb{K}_\omega^{(\sigma)} + \frac{v}{2\pi N l} \left[ (\delta \, \mathbb{K}_\omega^{(\sigma)}) \, \mathbb{K}_\omega + \mathbb{K}_\omega^{(\sigma)} \delta \, \mathbb{K}_\omega \right] . \tag{17.34}$$

If the expectation value with respect to $\Delta(r)$ is formed with the help of Eqs. (17.7) and (17.8), we obtain

$$(\Delta, \delta \, \mathbb{K}_\omega \Delta) = \frac{(\Delta, \delta \, \mathbb{K}_\omega^{(\sigma)} \Delta)}{\left( 1 - \dfrac{\kappa_\omega^{(\sigma)}}{l} \right)^2} . \tag{17.35}$$

Also Eq. (17.9) was used in the derivation of this result. Inserting Eq. (17.35) into Eq. (17.32), we derive

$$T \sum_\omega \frac{(\Delta, \delta \, \mathbb{K}_\omega^{(\sigma)} \Delta)}{\left( 1 - \dfrac{\kappa_\omega^{(\sigma)}}{l} \right)^2} = 0 \tag{17.36}$$

or the analogous relation following from Eq. (17.33).

The expectation value $(\Delta, \delta \, \mathbb{K}_\omega^{(\sigma)} \Delta)$ is now calculated with the help of essentially the same methods as were used in Section 15 b for the calculation of contributions to $\Phi$. Since only lowest order contributions shall be retained, we derive

$$\frac{v}{2\pi N} (\Delta, \delta \, \mathbb{K}_\omega^{(\sigma)} \Delta) = - \frac{\zeta_\omega^3}{3} \delta \lambda \int |\Delta(r)|^2 \, d^3 r + \frac{\zeta_\omega^4}{16} \oint p |n \times \tilde{\partial} \, \Delta(r)|^2 \, |d^2 r| \tag{17.37}$$

for a sample bounded by a non-specularly reflecting surface. The second term on the righthand side corresponds to Eq. (15.29). The first term is obtained since also $\lambda$, in the original expression (17.17), has to be altered in order that Eq. (17.36) is satisfied. In most cases of practical interest, the magnetic field is homogeneous. The quantity to be determined is the strength of the critical field for a given geometry of field and sample. Eq. (17.16), with the boundary condition (15.1), then defines $\lambda = \lambda(H)$. In this case,

$$\delta\lambda = \frac{d\lambda(H)}{dH}\,\delta H \tag{17.38}$$

is inserted into Eq. (17.37).

Eq. (17.36) may now be solved for the alteration of the critical field,

$$\delta H = \frac{\xi_0(T)}{d\lambda/dH}\,F_1\left(\frac{\xi_0(T)}{l}, \lambda\zeta_0^2(T)\right)\frac{\oint p|n \times \tilde{\partial}\Delta(r)|^2\,|d^2 r|}{\int |\Delta(r)|^2\,d^3 r}. \tag{17.39}$$

There is an additional term on the righthand side if Eq. (17.33) has to be used instead of Eq. (17.32). The function $F_1\left(\frac{\xi_0}{l}, \lambda\zeta_0^2\right)$ is defined by

$$F_1\left(\frac{\xi_0}{l}, \lambda\zeta_0^2\right) := \frac{3}{16\xi_0}\frac{\displaystyle\sum_\omega \frac{\zeta_\omega^2}{\left(2|\omega| + \frac{\lambda\zeta_\omega^2 v}{3l}\right)^2}}{\displaystyle\sum_\omega \frac{\zeta_\omega}{\left(2|\omega| + \frac{\lambda\zeta_\omega^2 v}{3l}\right)^2}}. \tag{17.40}$$

The lengthes $\xi_0(T)$ and $\zeta_0(T)$ were given in Eqs. (13.11) and (17.20), respectively.

The terms containing $\lambda$ may be neglected in the GL approximation when Eq. (17.28) is valid. They cancel in the dirty limit when $\zeta_\omega$ may be replaced by $l$ in the respective numerators. Therefore, $F_1\left(\frac{\xi_0}{l}, \lambda\zeta_0^2\right)$ may be replaced by the simpler function

$$F_1\left(\frac{\xi_0}{l}\right) := \frac{3}{16\xi_0}\frac{\displaystyle\sum_\omega \frac{\zeta_\omega^2}{(2|\omega|)^2}}{\displaystyle\sum_\omega \frac{\zeta_\omega}{(2|\omega|)^2}} \tag{17.41}$$

in all cases of practical interest (GL approximation and diffusion approximation). It is

$$F_1(\varrho) = \frac{45\,\zeta(4)}{224\,\zeta(3)} \tag{17.42}$$

in the clean limit ($\varrho \ll 1$) and

$$F_1(\varrho) = \frac{3}{16\varrho} \tag{17.43}$$

in the dirty limit ($\varrho \gg 1$). The function $F_1(\varrho)$ is represented graphically in Fig. 3. A useful interpolation between the two limiting cases is given by

$$F_1(\varrho) \approx \frac{1}{\dfrac{224\,\zeta(3)}{45\,\zeta(4)} + \dfrac{16\varrho}{3}} = \frac{0.1875}{1.037 + \varrho}. \tag{17.44}$$

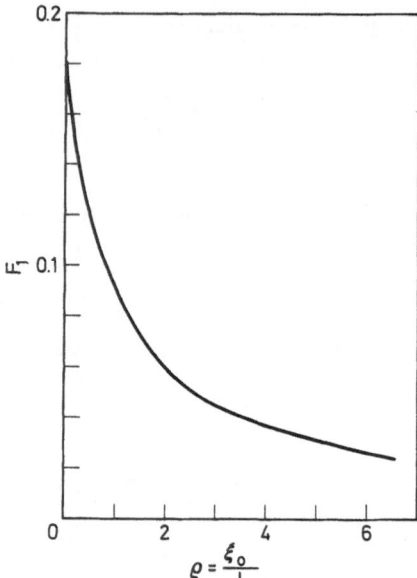

Fig. 3. The function $F_1(\xi_0/l)$ entering into Eq. (17.39) (for $\lambda = 0$) and defined in Eq. (17.41). The result of the interpolation formula (17.44) lies slightly above the curve, but it would not be discernible in the scale used in the figure

Incidentally it is

$$F_1(\varrho) = - \frac{3}{16\chi(\varrho)} \; \frac{d\chi(\varrho)}{d\varrho} \tag{17.45}$$

with $\chi(\varrho)$ from Eq. (13.28). $\xi_0$ may always be put equal to $\xi_0(T_c)$ in the GL approximation.

If the sample is bounded by a specularly reflecting surface ($p \equiv 0$), Eq. (17.39) does not lead to an alteration of the critical field. We have to go to a higher order and to include terms corresponding to Eq. (15.30) (with $p' = 0$ because of specular reflexion). It turns out in the applications (Chapter III), that it is now necessary to include the fourth derivative on the righthand side of Eq. (17.14). Eqs. (13.48) and (13.55) have to be remembered. In this way, Eq. (17.37) is replaced by

$$\frac{v}{2\pi N} (\varDelta, \delta \mathbb{K}_\omega^{(\sigma)} \varDelta) = \left[ - \frac{\zeta_\omega^3}{3} \frac{\partial \lambda}{\partial H} \delta H + \frac{\zeta_\omega^5}{5} (\lambda^2 + (2eH)^2) \right] \int |\varDelta(r)|^2 \, d^3 r$$

$$- \frac{\zeta_\omega^5}{15} \Big[ 2ie \oint \left( 3\varDelta^*(r) \, (d^2 r \times H) \cdot \tilde{\partial} \varDelta(r) + (\tilde{\partial} \varDelta(r))^* \cdot (d^2 r \times H) \, \varDelta(r) \right)$$

$$+ 2 \oint \frac{dn_i}{dx_l} (\tilde{\partial}_l \varDelta(r))^* \, \tilde{\partial}_i \varDelta(r) \, |d^2 r| \Big], \tag{17.46}$$

where the vectorial surface element, $d^2r$, is parallel to the normal unit vector, $n$, pointing towards the exterior. The last integral on the righthand side contains a derivative of this vector. It vanishes on a plane surface.

Eq. (17.36) is now solved for $\delta H$ with the result

$$\delta H = \frac{\zeta_0^2(T)}{d\lambda/dH} \, F_2\left(\frac{\zeta_0(T)}{l}, \lambda\zeta_0^2(T)\right)\left[3(\lambda^2 + (2eH)^2)\right.$$

$$\left. - \frac{2ie\oint(d^2r \times H)\cdot\left(3\Delta^*(r)\tilde{\partial}\Delta(r) + (\tilde{\partial}\Delta(r))^*\Delta(r)\right) + 2\oint\frac{\partial n_i}{\partial x_l}(\tilde{\partial}_l\Delta(r))^*\tilde{\partial}_i\Delta(r)|d^2r|}{\int|\Delta(r)|^2\,d^3r}\right]. \tag{17.47}$$

Again, an additional term appears on the righthand side if Eq. (17.33) is used instead of Eq, (17.32). The function $F_2\left(\frac{\zeta_0}{l}, \lambda\zeta_0^2\right)$ is defined by

$$F_2\left(\frac{\zeta_0}{l}, \lambda\zeta_0^2\right) := \frac{1}{5\zeta_0^2} \, \frac{\displaystyle\sum_\omega \frac{\zeta_\omega^3}{\left(2|\omega| + \frac{\lambda\zeta_\omega^2 v}{3l}\right)^2}}{\displaystyle\sum_\omega \frac{\zeta_\omega}{\left(2|\omega| + \frac{\lambda\zeta_\omega^2 v}{3l}\right)^2}}. \tag{17.48}$$

The terms containing $\lambda$ may again be neglected in the GL approximation. They cancel in the dirty limit. Again, $F_2\left(\frac{\zeta_0}{l}, \lambda\zeta_0^2\right)$ may be replaced by

$$F_2\left(\frac{\zeta_0}{l}\right) := \frac{1}{5\zeta_0^2} \, \frac{\displaystyle\sum_\omega \frac{\zeta_\omega^3}{(2|\omega|)^2}}{\displaystyle\sum_\omega \frac{\zeta_\omega}{(2|\omega|)^2}} \tag{17.49}$$

both in the GL approximation and in the diffusion approximation. It is

$$F_2(\varrho) = \frac{31\zeta(5)}{140\zeta(3)} \tag{17.50}$$

in the clean limit ($\varrho \ll 1$) and

$$F_2(\varrho) = \frac{1}{5\varrho^2} \tag{17.51}$$

in the dirty limit ($\varrho \gg 1$). The function $F_2(\varrho)$ is represented graphically in Fig. 4. A useful interpolation between the two limiting cases is given by

$$F_2(\varrho) \approx \frac{1}{\left(\sqrt{\frac{140\zeta(3)}{31\zeta(5)}} + \sqrt{5}\varrho\right)^2} = \frac{0.200}{(1.023 + \varrho)^2}. \tag{17.52}$$

It is

$$F_2(\varrho) = \frac{1}{10\chi(\varrho)} \, \frac{d^2\chi(\varrho)}{d\varrho^2} \tag{17.53}$$

with $\chi(\varrho)$ given in Eq. (13.28).

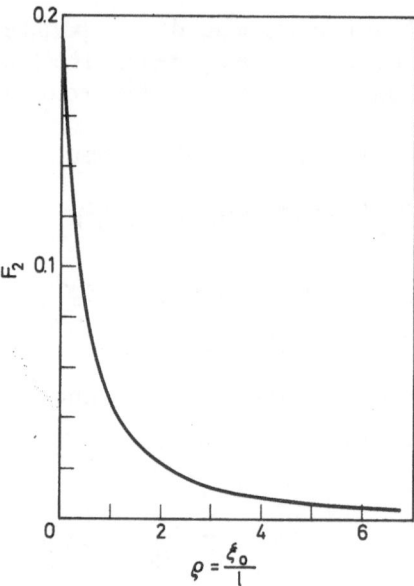

Fig. 4. The function $F_2(\zeta_0/l)$ entering into Eq. (17.47) (for $\lambda = 0$) and defined in Eq. (17.49). The result of the interpolation formula (17.52) lies slightly above the curve, but it would not be discernible in the scale used in the figure

In the dirty limit when the diffusion approximation holds, it is possible to go beyond the above approximations. The surface term contained in Eq. (17.37) may be incorporated into the original $(\varDelta, \mathbb{K}_\omega^{(\sigma)} \varDelta)$, since it becomes independent of $\omega$ if $\zeta_\omega$ is replaced by $l$. Now, a function $\varDelta(r)$, satisfying Eq. (13.1) and the boundary condition

$$\boldsymbol{n} \cdot \tilde{\partial} \varDelta(r) = -\frac{3l}{16} \left[ p(\boldsymbol{n} \times \tilde{\partial}) \cdot (\boldsymbol{n} \times \tilde{\partial}) \varDelta(r) + (\operatorname{grad} p) \cdot \tilde{\partial} \varDelta(r) \right], \quad (17.54)$$

is a common eigenfunction to all kernels $\mathbb{K}_\omega^{(\sigma)}$ and so to all $\mathbb{K}_\omega$. It is then a consistent approximation to include terms containing $p'$ on the righthand side of Eq. (17.46). These terms may be inferred from Eq. (15.30). Eq. (17.47) is so replaced by

$$\delta_1 H = \frac{l^2}{5 \, d\lambda/dH} \left[ 3(\lambda^2 + (2eH)^2) - \frac{1}{\int |\varDelta(r)|^2 \, d^3r} \left\{ 2ie \oint (d^2 r \times H) \right. \right.$$
$$\cdot \left[ \left( 3 + \frac{p'}{2} \right) \varDelta^*(r) \, \tilde{\partial} \varDelta(r) + \left( 1 - \frac{p'}{2} \right) (\tilde{\partial} \varDelta(r))^* \, \varDelta(r) \right] \qquad (17.55)$$
$$\left. \left. + 2 \oint \frac{\partial n_i}{\partial x_l} \left[ \left( 1 - \frac{p'}{2} \right) (\tilde{\partial}_l \varDelta(r))^* \, \tilde{\partial}_i \varDelta(r) - \frac{p'}{2} (\tilde{\partial}_i \varDelta(r))^* \, \tilde{\partial}_l \varDelta(r) \right] |d^2 r| \right\} \right].$$

A more systematic approximation is obtained in the following manner, where it is assumed that the diffuseness, $p$, is constant on the surface. Let $\lambda(H, p)$ be the lowest eigenvalue for a function satisfying Eq. (17.16) with boundary condition (17.54). We start from the magnetic field determined with boundary condition (15.1) ($p = 0$). The alteration of the magnetic field is then given by

$$\delta H = \delta_1 H + \delta_2 H, \tag{17.56}$$

where $\delta_1 H$ is calculated from Eq. (17.55) with

$$\lambda = \lambda(H, 0), \qquad \frac{d\lambda}{dH} = \frac{\partial \lambda(H, 0)}{\partial H}. \tag{17.57}$$

$\delta_2 H$ represents the alteration of $H$ due to $p \neq 0$. From

$$\lambda(H, 0) = \lambda(H + \delta_2 H, p), \tag{17.58}$$

it follows to the required approximation as

$$\delta_2 H = -p \frac{\dfrac{\partial \lambda}{\partial p}}{\dfrac{\partial \lambda}{\partial H}} - \frac{p^2}{2 \dfrac{\partial \lambda}{\partial H}} \left[ \left( \frac{\dfrac{\partial \lambda}{\partial p}}{\dfrac{\partial \lambda}{\partial H}} \right)^2 \frac{\partial^2 \lambda}{\partial H^2} - 2 \frac{\dfrac{\partial \lambda}{\partial p}}{\dfrac{\partial \lambda}{\partial H}} \frac{\partial^2 \lambda}{\partial H \partial p} + \frac{\partial^2 \lambda}{\partial p^2} \right], \tag{17.59}$$

with the derivatives taken at $H = 0$ and $p = 0$.

It is not clear whether it would be justified to derive local boundary conditions from the approximately evaluated functional (17.30). If this could, however, be done, the boundary condition would read

$$\boldsymbol{n} \cdot \tilde{\partial} \varDelta(\boldsymbol{r}) = -\xi_0(T) F_1\left(\frac{\xi_0}{l}\right) [p(\boldsymbol{n} \times \tilde{\partial}) \cdot (\boldsymbol{n} \times \tilde{\partial}) \varDelta(\boldsymbol{r}) + (\mathrm{grad}\, p) \cdot \tilde{\partial} \varDelta(\boldsymbol{r})] \tag{17.60}$$

with the function $F_1(\xi_0/l)$ defined by Eq. (17.41). Eq. (15.32), in the clean limit, and Eq. (17.54), in the dirty limit, are special cases of Eq. (17.60). Local boundary conditions are certainly well justified in these two limiting cases.

## 18. Electric Current Density

### a) General Formulations

At a second kind transition between superconducting and normal state, the electric current density in the superconductor becomes infinitesimally small. It does not produce an "internal" magnetic field which would modify the external field. The magnetic field acting upon the electrons is just the external field as is assumed everywhere in this review article. It is, nevertheless, not without interest to calculate even the infinitesimally small electric current density. The current pattern permits to conclude in which way the magnetic field will be modified by the current if one does not work exactly at the phase transition.

The calculation of the electric current density [5] at a second kind transition is based upon Eq. (2.24) or, in a more compact notation, upon

$$i(r') = T \sum_{\omega}' \left( \varDelta, \frac{\delta \mathbb{K}_\omega}{\delta A(r')} \varDelta \right). \tag{18.1}$$

Introducing the distribution function $\varDelta_\omega(r, v)$ from Eq. (11.22), we obtain

$$i(r') = 2\pi^2 \int N(r) \varDelta^*(r) T \sum_{\omega}' \frac{\delta \varDelta_\omega(r, v)}{\delta A(r')} d^3 r d\Omega. \tag{18.2}$$

A Boltzmann equation for the functional derivative is obtained from the Boltzmann equation (11.23), if use is made of Eq. (2.27). The result is

$$(2|\omega| + v \cdot \tilde{\partial} + nv\sigma) \frac{\delta \varDelta_\omega(r, v)}{\delta A(r')} - nv \oint \frac{d\sigma(\theta)}{d\Omega} \frac{\delta \varDelta_\omega(r, v')}{\delta A(r')} d\Omega' \tag{18.3}$$
$$= -2iev \varDelta_\omega(r, v) \delta(r - r').$$

The functional derivative obeys the same boundary conditions on surfaces and interfaces as the distribution function, $\varDelta_\omega(r, v)$, itself.

Eq. (18.3) is solved with the help of a classical Green's function, $e_\omega(r, v; r'', v'')$, satisfying

$$(2|\omega| + v \cdot \tilde{\partial} + nv\sigma) e_\omega(r, v; r'', v'') - nv \oint \frac{d\sigma(\theta)}{d\Omega} e_\omega(r, v'; r'', v'') d\Omega' \tag{18.4}$$
$$= \delta(r - r'') \hat{\delta}(v, v'')$$

and appropriate boundary conditions. Evidently, it is

$$\varDelta_\omega(r, v) = \frac{1}{(2\pi)^2} \int e_\omega(r, v; r', v') \varDelta(r') d^3 r' d\Omega'. \tag{18.5}$$

Further, the solution to Eq. (18.3) plus boundary conditions is given by

$$\frac{\delta \varDelta_\omega(r, v)}{\delta A(r')} = -2ie \oint e_\omega(r, v; r', v'') v'' \varDelta_\omega(r', v'') d\Omega''. \tag{18.6}$$

Inserting this equation into Eq. (18.2) and making use of Eq. (16.11), which is also valid for $e_\omega(r, v; r', v')$, and of Eq. (18.5), we finally arrive at the general expression

$$i(r) = -ie(2\pi)^4 N(r) T \sum_{\omega}' \oint \varDelta_\omega^*(r, -v) v \varDelta_\omega(r, v) d\Omega \tag{18.7}$$

for the current density. Eq. (18.7) may, with the help of Eq. (11.22), also be written in the form

$$i(r) = -ie(2\pi)^4 N(r) T \sum_{\omega}' \int \varDelta^*(r') g_\omega^*(r, -v; r') v g_\omega(r, v; r'') \varDelta(r'') d^3 r' d^3 r'' d\Omega. \tag{18.8}$$

The current density becomes indeed infinitesimally small at a second kind transition ($\Delta(r) \to 0$). It might be noticed that there is no simple relation between the electric current density and the flux of electrons (or carriers of information) as introduced by Eq. (14.57). The electric current density vanishes if the distribution function, $\Delta_\omega(r, v)$, is real (or real apart from a complex factor which at most depends upon $r$).

The distribution function, $g_\omega(r, v; r')$, is real in the absence of a magnetic field [$A(r) \equiv 0$]. Also $\Delta(r)$ may be chosen real since, if it were not, $\Delta(r)$, $\Delta^*(r)$, and their sum would satisfy the gap equation (10.1). Therefore, there is no electric current flowing in the superconductor in the absence of an external magnetic field [more correctly: $\Delta(r)$ can always be chosen such that there is no current; but no solution $\Delta(r)$, characteristic for a second kind transition in the absence of a magnetic field, has so far been encountered which corresponds to a current].

It is easily seen that the electric current density (18.7) satisfies all conditions to be postulated for a stationary current density. It is a gauge invariant real quantity. Its divergence vanishes provided that the gap equation (10.1) is satisfied. It indeed follows from the Boltzmann equation (11.23) that it is

$$\operatorname{div} i(r) = i e (2\pi)^2 N(r) T \sum_\omega{}' \oint (\Delta^*(r) \Delta_\omega(r, v) - \Delta_\omega^*(r, -v) \Delta(r)) \, d\Omega. \qquad (18.9)$$

The righthand side is shown to vanish if Eq. (10.1) is taken into account. Further, the normal component of the current density vanishes on external surfaces,

$$n \cdot i(r) = 0, \qquad (18.10)$$

because of the condition (10.19) to be satisfied by the reflexion kernel. Its normal component is continuous on interfaces since the reflexion and transmission kernels satisfy Eqs. (10.29) and (10.30).

It might be noticed that the general conditions on reflexion and transmission kernels, of relevance in the present context, are consequences of the invariance under time reversal. That does not mean that there is a mysterious connection between electric current density and time reversal invariance. But time reversal invariance [i. e., Eq. (16.11)] entered already into the step leading from Eq. (18.6) to Eq. (18.7).

## b) Approximations

The electric current density corresponding to various approximations is obtained from Eq. (18.7) by inserting approximate expressions for the distribution function, $\Delta_\omega(r, v)$. It has then to be checked whether the vanishing of the divergence and the continuity of the normal component are also guaranteed in the approximation under discussion.

In the GL approximation, we take the distribution function from Eq. (15.2). With the help of Eq. (13.6), we obtain

$$\boldsymbol{i}(\boldsymbol{r})$$

$$= i e \frac{4\pi}{3} N(\boldsymbol{r}) v^2(\boldsymbol{r}) T \sum_\omega \frac{1}{(2|\omega|)^2 (2|\omega| + n v \sigma_{\mathrm{tr}})} \left[ \varDelta^*(\boldsymbol{r}) \tilde{\partial} \varDelta(\boldsymbol{r}) - (\tilde{\partial} \varDelta(\boldsymbol{r}))^* \varDelta(\boldsymbol{r}) \right], \tag{18.11}$$

where the cut-off may again be neglected in the weak-coupling case. Essentially this expression with the typically quantum-mechanical current density was already proposed by *Ginzburg* and *Landau* in their original paper [8]. The microscopic derivation of Eq. (18.11) was first given by *Gorkov* [40, 41]. This current density is not only a gauge invariant real quantity. It further satisfies

$$\operatorname{div} \boldsymbol{i}(\boldsymbol{r}) = 0 \tag{18.12}$$

as a consequence of the GL equation (13.1). This is also true for a position dependent impurity concentration when Eq. (13.35) has to be used. For methodical reasons it should be remembered that the gap equation (10.1), which was decisive for the vanishing of the righthand side of Eq. (18.9), already entered into the derivation of the GL equation. Eq. (18.10) follows on an external surface from the boundary condition (15.1). But this boundary condition can, of course, not be derived from the requirement (18.10) The continuity of the normal component of $\boldsymbol{i}(\boldsymbol{r})$ on interfaces follows as well from the boundary conditions (15.62) as from the simplified boundary conditions (15.67).

Close to surfaces and interfaces, actually Eq. (15.6) ought to be used whereas Eq. (18.11) is obtained by only inserting $\varDelta_{\omega 0}(\boldsymbol{r}, \boldsymbol{v})$ into Eq. (18.7). The additional terms obtained by making use of the complete Eq. (15.6) could lead to additional currents flowing in the surface layers. That already the normal component of the current density calculated from $\varDelta_{\omega 0}(\boldsymbol{r}, \boldsymbol{v})$ vanishes on surfaces and is continuous on interfaces, is a strong argument in favour of the absence of such surface and interface currents. They appear, however, in higher orders when boundary condition (15.1) is replaced by boundary condition (15.32): Eq. (18.10) is in general no longer satisfied by the $\varDelta_{\omega 0}(\boldsymbol{r}, \boldsymbol{v})$ current density (18.11) alone. That means that there now exists a surface current to which $\varDelta_{\omega 1}(\boldsymbol{r}, \boldsymbol{v})$ contributes.

In the diffusion approximation, the distribution function $\varDelta_\omega(\boldsymbol{r}, \boldsymbol{v})$ is given by Eq. (14.1). Inserting it into Eq. (18.7), we obtain

$$\boldsymbol{i}(\boldsymbol{r}) = - \tfrac{2}{3} i e (2\pi)^5 N v^2 T \sum_\omega (H_\omega^*(\boldsymbol{r}) H_\omega(\boldsymbol{r}) - H_\omega^*(\boldsymbol{r}) H_\omega(\boldsymbol{r})). \tag{18.13}$$

Application of Eq. (14.6) leads to

$$\boldsymbol{i}(\boldsymbol{r}) = \tfrac{2}{3} i e (2\pi)^5 N v l_{\mathrm{tr}} T \sum_\omega [H_\omega^*(\boldsymbol{r}) \tilde{\partial} H_\omega(\boldsymbol{r}) - (\tilde{\partial} H_\omega(\boldsymbol{r}))^* H_\omega(\boldsymbol{r})], \tag{18.14}$$

which again shows a typically quantum-mechanical structure. The right-hand side of Eq. (18.14) is again real and gauge invariant. Eq. (18.12) is a consequence of the diffusion equation (14.8), if also Eq. (14.9) is taken into account. Eq. (18.10) follows from the boundary condition (16.2). The continuity of the normal component of the current density on interfaces follows from the boundary conditions (16.36) or from the simplified boundary conditions (16.38). It seems that there are again no surface and interface currents.

Eq. (18.14) may also be written in the form

$$i(r) = \frac{2}{3} ie \frac{v l_{tr}}{2\pi N} T \sum_\omega \int \Delta^*(r') [K_\omega^*(r, r') \tilde{\partial} K_\omega(r, r'') - (\tilde{\partial} K_\omega(r, r'))^* K_\omega(r, r'')] \Delta(r'') d^3r' d^3r'',$$

(18.15)

where $K_\omega(r, r')$ is the approximate kernel satisfying the diffusion equation (14.10), the boundary condition (16.3) on external surfaces, and the boundary conditions (16.39) or (16.40) on interfaces. Eq. (18.15) may also be derived directly from Eq. (2.24) since

$$\frac{\delta K_\omega(r', r'')}{\delta A(r)} = \frac{2}{3} ie \frac{v l_{tr}}{2\pi N} [K_\omega^*(r, r') \tilde{\partial} K_\omega(r, r'') - (\tilde{\partial} K_\omega(r, r'))^* K_\omega(r, r'')] \quad (18.16)$$

follows from the diffusion equation (14.10) and its variational derivative,

$$\left(2|\omega| - \frac{v l_{tr}}{3} \tilde{\partial} \cdot \tilde{\partial}\right) \frac{\delta K_\omega(r, r')}{\delta A(r'')} = 2ie \frac{v l_{tr}}{3} [\delta(r - r'') \tilde{\partial} + \tilde{\partial} \delta(r - r'')] K_\omega(r, r'). \quad (18.17)$$

Also the boundary conditions and Eq. (11.16) have to be used.

It was shown in Section 14c, that a GL like formulation may be given to the diffusion approximation if the sample consists of a single metal. It follows from Eq. (18.14) and from

$$H_\omega(r) = \frac{1}{(2\pi)^2} \frac{1}{2|\omega| + \frac{\lambda v l_{tr}}{3}} \Delta(r), \quad (18.18)$$

that it is

$$i(r) = \frac{2}{3} ie 2\pi N v l_{tr} T \sum_\omega \frac{1}{\left(2|\omega| + \frac{\lambda v l_{tr}}{3}\right)^2} [\Delta^*(r) \tilde{\partial} \Delta(r) - (\tilde{\partial} \Delta(r))^* \Delta(r)]. \quad (18.19)$$

The $\Delta$ dependent part is the same as in Eq. (18.11), whereas the other factors are not of great practical importance. Eqs. (18.11) and (18.19) become strictly identical in the region of common validity of the GL approximation ($T \approx T_c$) and the diffusion approximation [Eq. (14.15)].

A special approximation scheme was presented in Section 17. It is applicable to samples consisting of a single metal with homogeneous impurity concentration and isotropic scattering cross section. GL approximation and diffusion approximation were contained in this scheme as special cases. The current density shall now be derived for the approximation of Section 17b. First it follows from Eq. (17.35) that it is

$$
\left( \Delta, \frac{\delta \, \mathbb{K}_\omega}{\delta A(r)} \, \Delta \right) = \frac{\left( \Delta, \dfrac{\delta \, \mathbb{K}_\omega^{(\sigma)}}{\delta A(r)} \, \Delta \right)}{\left( 1 - \dfrac{\kappa_\omega^{(\sigma)}}{l} \right)^2}.
\tag{18.20}
$$

In the denominator, Eq. (17.17) has to be inserted whereas the current density corresponding to $\left( \Delta, \dfrac{\delta \, \mathbb{K}_\omega^{(\sigma)}}{\delta A(r)} \, \Delta \right)$ follows from the GL approximation result (18.11) for a clean metal ($n = 0$) with the substitution $2|\omega| \to 2|\omega| + nv\sigma$. In this way we arrive at

$$
i(r)
$$
$$
= \frac{2}{3} i e \, 2\pi N v^2 \, T \sum_\omega \frac{1}{(2|\omega| + nv\sigma)\left( 2|\omega| + \dfrac{\lambda v \zeta_\omega^2}{3l} \right)^2} \, [\Delta^*(r)\tilde{\partial}\Delta(r) - (\tilde{\partial}\Delta(r))^* \Delta(r)]
\tag{18.21}
$$

with again the typical $\Delta$ dependent part.

# III. Applications

## 19. The Bulk Critical Field $H_{c2}$

Let us imagine an infinitely extended, macroscopically homogeneous metal in a homogeneous magnetic field. A transition between normal and superconducting state may be either of the first or of the second kind. If the transition is of the first kind, the metal is called a type one superconductor. The critical field is denoted by $H_c(T)$. If the transition is of the second kind, the metal is called a type two superconductor. The critical field is denoted by $H_{c2}(T)$. Both critical fields are in principle well-defined quantities in a reversible superconductor. Whether the superconductor is type one or type two, depends upon whether $H_c(T)$ is bigger or smaller than $H_{c2}(T)$.

### a) Ginzburg-Landau and Diffusion Approximation

The bulk critical field $H_{c2}$ was discovered theoretically by *Abrikosov* [47] on the basis of the GL theory. In order to reproduce his analysis we assume a homogeneous magnetic field, with strength $H$, parallel to the $z$ direction. The components of the vector potential are chosen as

$$A_x = A_z = 0, \qquad A_y = H x. \tag{19.1}$$

The linearized GL equation (13.1) reads

$$\left[\frac{\partial^2}{\partial x^2} + \left(\frac{\partial}{\partial y} + 2ieHx\right)^2 + \frac{\partial^2}{\partial z^2}\right] \Delta(r) = -\lambda(T)\,\Delta(r) \tag{19.2}$$

where $\lambda(T)$ is either given by Eq. (13.26) (GL approximation) or by Eq. (14.31) (diffusion approximation). In the infinite sample, the solution, $\Delta(r)$, to Eq. (19.2) has to be bounded at infinity. The ansatz

$$\Delta(r) = f(x) \exp[i(ky + \mu z)] \tag{19.3}$$

suggests itself. The parameter $\mu$ may, in contrast to $k$, also be defined in a gauge invariant manner,

$$H \cdot \tilde{\partial} \Delta(r) = iH\mu \Delta(r). \tag{19.4}$$

The function $f(x)$ obeys the differential equation

$$\left[\frac{d^2}{dx^2} - (k + 2eHx)^2\right] f(x) = [\mu^2 - \lambda(T)]\, f(x). \tag{19.5}$$

The mathematical structure of the lefthand side is known from the quantum mechanical treatment of the harmonic oscillator. The solutions bounded at $x \to \pm \infty$ actually approach zero in these limits. It is

$$\left[\frac{d^2}{dx^2} - (k + 2eHx)^2\right] f(x) = -4eH\left(n + \frac{1}{2}\right) f(x) \quad (n = 0, 1, 2, \ldots)$$
(19.6)

or

$$4eH\left(n + \frac{1}{2}\right) = \lambda(T) - \mu^2.$$
(19.7)

The magnetic field assumes its maximal value and so becomes equal to the critical field, $H_{c2}$, for $n = \mu = 0$,

$$2eH_{c2}(T) = \lambda(T).$$
(19.8)

If $\hbar$ and $c$ are no longer put equal to one and if $\xi(T)$ is introduced from Eq. (13.2), Eq. (19.8) may be rewritten* as

$$2\pi H_{c2}(T)\, \xi^2(T) = \Phi_0,$$
(19.9)

where $\Phi_0 = hc/2e = 2.07 \cdot 10^{-7} \text{G cm}^2$ is the so-called unit of quantized flux. For the sake of future reference, the pair potential shall also be written down explicitly,

$$\Delta(r) = \exp\left[-eH\left(x + \frac{k}{2eH}\right)^2 + iky\right].$$
(19.10)

The electric current density is, up to a real constant factor, given by

$$i(r) \propto i\,[\Delta^*(r)\,\tilde{\partial}\,\Delta(r) - (\tilde{\partial}\Delta(r))^*\,\Delta(r)]\,;$$
(19.11)

cf. Eqs. (18.11) and (18.19). Insertion of the solution (19.10), with $\mu$ put equal to zero, leads to

$$i_x = i_z = 0, \quad i_y \propto -2(k + 2eHx)\left|f\left(x + \frac{k}{2eH}\right)\right|^2,$$
(19.12)

where $f(x)$ is given by

$$f(x) = \exp(-eHx^2).$$
(19.13)

There is no net current through the sample since it is

$$\int_{-\infty}^{+\infty} i_y\, dx = 0,$$
(19.14)

which follows from the fact that $|f(x)|^2$ is an even function.

----

* The formal trick is always, to replace the unit of charge, $e$, by $\pi/\Phi_0$. Relations like Eq. (19.9) are correct if the magnetic field (or, rather, the magnetic induction) is measured in Gaussian units (Gauss).

Eq. (19.5) shows a degeneracy with respect to $k$. Any linear combination of solutions (19.10) is again an admissable solution. Which one is the correct solution cannot be decided within the scope of the linear theory. The periodic Abrikosov pattern [47] is produced by the linear combination

$$\Delta(r) = \sum_{n=-\infty}^{+\infty} \exp\left[i(\alpha n^2 + k_0 y n)\right] f\left(x + \frac{nk_0}{2eH}\right) \tag{19.15}$$

with arbitrary constant parameters $\alpha$ and $k_0$ and with $f(x)$ given by Eq. (19.13). The right-hand side of Eq. (19.15) is periodic, apart from gauge factors, under the substitutions

$$x \to x, \quad y \to y + \frac{2\pi}{k_0} \tag{19.16}$$

and

$$x \to x + \frac{k_0}{2eH}, \quad y \to y + \frac{2\alpha}{k_0}, \tag{19.17}$$

which define a plane lattice. The area, $A$, of the unit cell of this lattice is given by

$$A = \frac{2\pi}{2eH_{c2}} = 2\pi \xi^2(T), \tag{19.18}$$

which means that the unit of quantized flux, $\Phi_0$, passes through the unit cell [cf. Eq. (19.9)]. The current pattern is strictly periodical with respect to the substitutions (19.16) and (19.17). There is still no net current flowing.

## b) A General Theorem

The gap equation (10.1) can only in certain limiting cases [essentially $T \approx T_c$ or $l_{tr} \ll \xi_0(T_c)$] be approximately replaced by the GL equation. In the $H_{c2}$ problem it is, nevertheless, true [48] under rather general conditions, that the complete set of common eigenfunctions to Eqs. (17.16) and (19.4) is also the set of eigenfunctions to each kernel, $\mathbb{K}_\omega$. Furthermore the eigenvalue depends only upon $\lambda$, $|\mu|$, and the field strength, $H$,

$$\mathbb{K}_\omega \Delta(r) = \frac{2\pi N}{v} \kappa_\omega(\lambda, |\mu|, H) \Delta(r). \tag{19.19}$$

The gap equation (10.1) reduces to Eq. (17.25), which may also be written in the form

$$\ln\left(\frac{T_c}{T}\right) = 2\pi T \sum_\omega{}' \left(\frac{1}{2|\omega|} - \frac{\kappa_\omega}{v}\right). \tag{19.20}$$

The general statements made above are not consequences of the method of correlation function. But since no explicit proofs seem to have been published, they shall be presented here. It follows from the behaviour of $K_\omega(r, r')$ under gauge tranformations [Eq. (3.2)], that

$$\tilde{K}_\omega(r, r') = \exp\left[2ie \int_{r'}^{r} A(s) \cdot ds\right] K_\omega(r, r'), \tag{19.21}$$

with integration along a straight line from $r'$ to $r$, is a gauge invariant function. It, therefore, depends only upon the magnetic field, $H$, itself and not upon the vector potential. Since the sample is, in the present case, infinitely extended and if it is macroscopically homogeneous and isotropic, $\tilde{K}_\omega(r, r')$ can depend only upon the translationally and rotationally invariant combinations $(r - r')^2$ and $H \cdot (r - r')$, and, of course, upon $H$. "Macroscopically homogeneous" means that the concentration of impurities (if any) is position independent. "Isotropic" means that the Fermi surface is spherical and that the differential scattering cross section (if impurities are present) depends only upon the scattering angle.

The above theorem is not valid for non-spherical Fermi surfaces. The $H_{c2}$ problem was for this case studied by *Hohenberg* and *Werthamer* [49]. The calculations were confirmed and extended with the help of the method of the correlation function by *Harms* [42]. *Harms* showed that *Hohenberg* and *Werthamer*'s calculation for $T = 0$ is unjustified.

In the spirit of Section 11, $K_\omega(r, r')$ shall now be regarded as the matrix element of a Hermitian operator, cf. Eq. (11.17). We further introduce another Hermitian operator, $\tilde{\mathbb{p}}$, by its matrix element,

$$\langle r|\tilde{\mathbb{p}}| r'\rangle = -i\tilde{\partial}\delta(r - r') = -i\left(\frac{\partial}{\partial r} + 2ieA(r)\right)\delta(r - r').$$  (19.22)

We want to show that $\mathbb{K}_\omega$ commutes with the mutually commuting Hermitian operators $\tilde{\mathbb{p}} \cdot \tilde{\mathbb{p}}$ and $H \cdot \tilde{\mathbb{p}}$. Therefore, these three operators possess a complete set of common eigenfunctions.

It is

$$\langle r|\tilde{\mathbb{p}}\,\mathbb{K}_\omega - \mathbb{K}_\omega\,\tilde{\mathbb{p}}| r'\rangle = \int(\langle r|\tilde{\mathbb{p}}| r''\rangle\langle r''|\mathbb{K}_\omega| r'\rangle - \langle r|\mathbb{K}_\omega| r''\rangle\langle r''|\tilde{\mathbb{p}}| r'\rangle)\,d^3 r''$$

$$= -i\left(\frac{\partial}{\partial r} + 2ieA(r) + \frac{\partial}{\partial r'} - 2ieA(r')\right)K_\omega(r, r').$$  (19.23)

Here, Eq. (19.21) is now inserted. Making use of

$$\left(\frac{\partial}{\partial r} + 2ieA(r)\right)\exp\left[-2ie\int_{r'}^{r}A(s)\cdot ds\right] = \left(\frac{\partial}{\partial r'} - 2ieA(r')\right)\exp\left[-2ie\int_{r'}^{r}A(s)\cdot ds\right]$$

$$= -ie(r - r')\times H\exp\left[-2ie\int_{r'}^{r}A(s)\cdot ds\right],$$  (19.24)

and of the translational invariance of $\tilde{K}_\omega(r, r')$,

$$\left(\frac{\partial}{\partial r} + \frac{\partial}{\partial r'}\right)\tilde{K}_\omega(r, r') = 0,$$  (19.25)

we obtain

$$\langle r|\tilde{\mathbb{p}}\,\mathbb{K}_\omega - \mathbb{K}_\omega\,\tilde{\mathbb{p}}| r'\rangle = -e(r - r')\times H K_\omega(r, r').$$  (19.26)

From this equation,

$$(H \cdot \tilde{\mathbb{p}})\,\mathbb{K}_\omega - \mathbb{K}_\omega(H \cdot \tilde{\mathbb{p}}) = 0$$  (19.27)

follows immediately, since the matrix element of the lefthand side vanishes identically.

The matrix element

$$\langle r|\tilde{\mathbb{p}}^2\,\mathbb{K}_\omega - \mathbb{K}_\omega\,\tilde{\mathbb{p}}^2| r'\rangle = \left[-\left(\frac{\partial}{\partial r} + 2ieA(r)\right)^2 + \left(\frac{\partial}{\partial r'} - 2ieA(r')\right)^2\right]K_\omega(r, r')$$  (19.28)

is evaluated in essentially the same way. Again Eq. (19.21) is inserted. Firstly, it is

$$\left(\frac{\partial}{\partial r} + 2ieA(r)\right)^2 \exp\left[-2ie\int_{r'}^{r} A(s)\cdot ds\right] = \left(\frac{\partial}{\partial r'} - 2ieA(r')\right)^2 \exp\left[-2ie\int_{r'}^{r} A(s)\cdot ds\right] \quad (19.29)$$

as a consequence of Eq. (19.24) and of

$$\frac{\partial}{\partial r} \times (r - r') = 0 = \frac{\partial}{\partial r'} \times (r - r'). \quad (19.30)$$

Secondly, it is

$$\left(-\frac{\partial^2}{\partial r^2} + \frac{\partial^2}{\partial r'^2}\right)\tilde{K}_\omega(r, r') = \left(-\frac{\partial}{\partial r} + \frac{\partial}{\partial r'}\right)\left(\frac{\partial}{\partial r} + \frac{\partial}{\partial r'}\right)\tilde{K}_\omega(r, r') = 0 \quad (19.31)$$

because of Eq. (19.25). Thirdly, it is

$$((r - r') \times H)\cdot \frac{\partial}{\partial r}\tilde{K}_\omega(r, r') = 0 = ((r - r') \times H)\cdot\frac{\partial}{\partial r'}\tilde{K}_\omega(r, r') \quad (19.32)$$

since $\tilde{K}_\omega(r, r')$ depends upon $r$ and upon $r'$ only through $(r - r')^2$ and $H\cdot(r - r')$, so that the gradient lies in the plane spanned by $r - r'$ and by $H$. In this way it is, finally, seen that the righthand side of Eq. (19.28) vanishes identically,

$$\hat{\mathbb{p}}^2\mathbb{K}_\omega - \mathbb{K}_\omega\hat{\mathbb{p}}^2 = 0. \quad (19.33)$$

The eigenvalue equations (17.16) and (19.4) do not specify the eigenfunctions completely. We want to show that this degeneracy is irrelevant for the present problem. This question shall be studied in the special gauge (19.1), which may be done without loss of generality. In this case, it is

$$\int_{r'}^{r} A(s)\cdot ds = \frac{H}{2}(y - y')(x + x'). \quad (19.34)$$

Since $\tilde{K}_\omega(r, r')$ depends only upon $r - r'$, again the ansatz (19.3) may be made. The degeneracy mentioned is connected with the parameter $k$. The ansatz (19.3) leads to an eigenvalue equation for the function $f(x)$. It is easily seen, that

$$\Delta(r) = f(x)\exp[i\mu z] \quad (19.35)$$

and

$$\Delta(r) = f\left(x + \frac{k}{2eH}\right)\exp[i(ky + \mu z)] \quad (19.36)$$

are eigenfunctions to Eq. (19.19) with the same eigenvalue which, therefore, does not depend upon $k$.

The eigenvalue $\kappa_\omega$ [Eq. (19.19)] does also not depend upon the sign of $\mu$. This fact follows from the relation

$$K_\omega(x, x', y - y', z - z') = K_\omega(x, x', y - y', -(z - z')) \quad (19.37)$$

which is valid in the special gauge (19.1). Eq. (19.37) may be proved on the basis of the general quantum mechanical definition (2.16), since the unique solution to Eq. (2.25) satisfies

$$G_\omega^{(0)}(x, x', y - y', z - z') = G_\omega^{(0)}(x, x', y - y', -(z - z')) \quad (19.38)$$

in the gauge (19.1). Eq. (19.37) should still be correct after a suitable average with respect to a distribution of impurities has been performed. Eq. (19.37) may also be proved with the help of the method of the correlation function from Eq. (10.4). The unique solution to the Boltzmann equation (10.5) satisfies

$$g_\omega(x, x', y - y', z - z', v_x, v_y, v_z) = g_\omega(x, x', y - y', -(z - z'), v_x, v_y, -v_z), \qquad (19.39)$$

from which relation Eq. (19.37) is indeed obtained.

The critical field $H_{c2}$ is, in the GL approximation and in the diffusion approximation, found by putting $\lambda$ in Eq. (17.16) equal to $2eH_{c2}$ and $\mu$ in Eq. (19.4) equal to zero. This was shown in Section 19a. It is usually assumed that this choice of eigenvalues and eigenfunctions is also appropriate in the general case. It may be conjectured that in this case the current density is still given by Eq. (19.11).

In order to produce arguments in favour of the validity of Eq. (19.11), we first notice that it is $i_z(r) = 0$ or

$$H \cdot i(r) = 0 \qquad (19.40)$$

for $\mu = 0$. This follows, on a quantum-mechanical basis, from Eqs. (2.17) and (19.38). It follows, on the basis of the method of the correlation function, from Eqs. (18.8) and (19.39). Now we try to express $i(r)$ in a bilinear way by $\Delta^*(r)$, $\Delta(r)$ and their gauge invariant derivatives. We have to form a real vector whose divergence vanishes as a consequence of Eqs. (17.16) and (19.4). We, therefore, put

$$i(r) = ia[\Delta^*(r)\,\tilde{\partial}\Delta(r) - (\tilde{\partial}\Delta(r))^*\,\Delta(r)] + bH|\Delta(r)|^2. \qquad (19.41)$$

This seems to be the most general expression, since higher order derivatives may be carried out with the help of Eqs. (17.16) and (19.4) and of Eq. (11.57). The real constants, $a$ and $b$, depend upon the eigenvalues, $\lambda$ and $\mu$, the field strength, $H$, and parameters of the metal (Fermi velocity, mean free path etc.). It is $b = 0$ for $\mu = 0$, since otherwise a contradiction to Eq. (19.40) would arise.

## c) Special Examples

The task of calculating the critical field $H_{c2}$ is very much simplified by the theorem of Sec. 19b. This theorem states the eigenfunctions to the operators $\mathbb{K}_\omega$ are known a priori. For $\lambda = 2eH, \mu = 0$, and the special (but irrelevant) choice $k = 0$, the pair potential follows from Eq. (19.10) as

$$\Delta(r) = \exp(-eHx^2) \qquad (19.42)$$

in the gauge (19.1). $H_{c2}$ is obtained as an implicit function of $T$, if the $H$ dependent eigenvalue $\kappa_\omega$ [Eq. (19.19)] is inserted into Eq. (19.20). Our main further task is to calculate $\kappa_\omega(H)$.

The calculation is particularly simple for a metal without impurities [48]. In this case, it is

$$K_\omega(r, r') = \frac{N}{2v|r - r'|^2} \exp\left[-\frac{|r - r'|}{\xi_\omega} - ieH(y - y')(x + x')\right], \qquad (19.43)$$

as follows from Eqs. (10.4), (12.5), (12.4), and (19.34). The eigenvalue $\kappa_\omega$ [Eq. (19.19)] is obviously given by

$$\kappa_\omega = \int \frac{1}{4\pi|r-r'|^2} \exp\left[-\frac{|r-r'|}{\xi_\omega} + eH(x-x'-i(y-y'))(x+x')\right] d^3r'.$$
(19.44)

The additional $x$ dependence turns out to be only apparent, since equivalent formulations are given by

$$\kappa_\omega = \int \frac{1}{4\pi r^2} \exp\left[-\frac{r}{\xi_\omega} - \frac{eH}{2}(x^2+y^2)\right] d^3r \qquad (19.45)$$

$(r = \sqrt{x^2+y^2+z^2})$ and

$$\kappa_\omega = \frac{1}{eH} \int_0^\infty \arctan(k\xi_\omega) \exp\left(-\frac{k^2}{2eH}\right) dk. \qquad (19.46)$$

Further substantial simplifications do not seem attainable.

Eq. (19.45) is derived from Eq. (19.44) if plane polar coordinates are introduced by

$$x'-x = \varrho\cos\varphi, \quad y'-y = \varrho\sin\varphi. \qquad (19.47)$$

If further the complex variable $\zeta = \exp(i\varphi)$ is introduced, the calculus of residues leads to

$$\int_0^{2\pi} \exp[eH(x-x'-i(y-y'))(x+x')] d\varphi = 2\pi \exp\left[-\frac{eH\varrho^2}{2}\right]. \qquad (19.48)$$

Eq. (19.46) is derived from Eq. (19.45) with the help of the Fourier representations

$$\frac{1}{r^2}\exp\left[-\frac{r}{\xi_\omega}\right] = \frac{1}{2\pi^2}\int \frac{\arctan(k\xi_\omega)}{k} \exp(ik\cdot r)\, d^3k,$$
$$\exp\left[-\frac{eH}{2}(x^2+y^2)\right] = \frac{1}{2\pi eH}\int \exp\left[-\frac{k_x^2+k_y^2}{2eH} + ik\cdot r\right]\delta(k_z)\, d^3k.$$
(19.49)

Inserting Eq. (19.49) into Eq. (19.45) and performing the integration with respect to $r$, we first obtain

$$\kappa_\omega = \frac{1}{2\pi eH} \int \frac{\arctan(k\xi_\omega)}{k} \exp\left[-\frac{k^2}{2eH}\right] dk_x\, dk_y \qquad (19.50)$$

with $k^2 = \sqrt{k_x^2 + k_y^2}$. Introduction of plane polar coordinates leads to Eq. (19.46).

The analysis is not valid down to arbitrarily small temperatures since, finally, such $\omega$ values will contribute to the sum on the righthand side of the gap equation (10.1) for which Eq. (7.22) is not fulfilled. The curvature of the orbits in the magnetic field may not be neglected for ·such small values of $|\omega|$ so that the method of the correlation function breaks down. The Landau quantization of the electric states becomes relevant. It leads to an oscillatory structure in the dependence of the transition temperature upon the magnetic field [50].

The previous analysis is easily extended [48] to a metal containing impurities of position independent concentration, which scatter isotropically. Operators $\mathbb{K}_\omega^{(\sigma)}$ and eigenvalues $\kappa_\omega^{(\sigma)}$ may be introduced with the help of Section 17a. $\kappa_\omega^{(\sigma)}(H)$ is obtained from Eqs. (19.44) through (19.46) by replacing the length $\xi_\omega$ by the length $\zeta_\omega$ [Eq. (12.14)]. Finally $\kappa_\omega$ is obtained from $\kappa_\omega^{(\sigma)}$ by means of Eq. (17.9).

It is more difficult to go beyond isotropic scattering. Only the simplest extension,

$$\frac{d\sigma(\theta)}{d\Omega} = \frac{\sigma}{4\pi}(1 + c\cos\theta), \tag{19.51}$$

appears to have been studied so far [51]. An analysis based upon the method of the correlation function [52] may start from the integral equation

$$\Delta_\omega(r, v) = \Delta_\omega^{(\sigma)}(r, v) + (2\pi)^2 \, nv \int d^3r' \, d\Omega' g_\omega^{(\sigma)}(r, v; r') \frac{d\sigma(\theta)}{d\Omega} \Delta_\omega(r', v'). \tag{19.52}$$

It follows from Eq. (12.16) and from the definition (11.22) of the distribution function $\Delta_\omega(r, v)$. $\Delta_\omega^{(\sigma)}(r, v)$ is defined in an analogous manner with $g_\omega(r, v; r')$ replaced by

$$g_\omega^{(\sigma)}(r, v; r')$$

$$= \frac{1}{(2\pi|r - r'|)^2 \, v} \exp\left[-\frac{|r - r'|}{\zeta_\omega} - ieH(y - y')(x + x')\right]\hat{\delta}(r - r', v); \tag{19.53}$$

cf. Eqs. (12.13) and (19.34).

Because of the uniqueness theorem of Section 11a, it suffices to conjecture an ansatz for $\Delta_\omega(r, v)$ and to show that it satisfies Eq. (19.52) with $\dfrac{d\sigma(\theta)}{d\Omega}$ given by Eq. (19.51). We put

$$\Delta_\omega(r, v) = a_\omega \, f_\omega(x, v) + b_\omega f'_\omega(x, v) \tag{19.54}$$

with

$$f_\omega(x, v) = \int \frac{1}{|r - r'|^2}$$

$$\cdot \exp\left[-\frac{|r - r'|}{\zeta_\omega} - ieH(y - y')(x + x') - eHx'^2\right]\hat{\delta}(r - r', v)\, d^3r',$$

$$f'_\omega(x, v) = \int \frac{x'[(x - x') - i(y - y')]}{|r - r'|^3} \tag{19.55}$$

$$\cdot \exp\left[-\frac{|r - r'|}{\zeta_\omega} - ieH(y - y')(x + x') - eHx'^2\right]\hat{\delta}(r - r', v)\, d^3r'$$

and constants, $a_\omega$ and $b_\omega$, to be determined below. The contribution $f_\omega(x, v)$ is suggested by the fact that it is

$$\Delta_\omega^{(\sigma)}(r, v) = \frac{1}{(2\pi)^2 \, v} f_\omega(x, v) \tag{19.56}$$

for the pair potential (19.42). It will further be seen that the contribution $f'_\omega(x, v)$ is produced by the non-isotropic term in Eq. (19.51).

Some auxiliary formulae are required. Integration over all directions of the velocity vector leads to

$$\oint f_\omega(x, v) \, d\Omega = f_\omega \Delta(r) , \quad \oint f'_\omega(x, v) \, d\Omega = f'_\omega \Delta(r) \tag{19.57}$$

with $\Delta(r)$ given by Eq. (19.42). The quantities $f_\omega$ and $f'_\omega$ do not depend upon $x$ and are given by

$$\begin{aligned}
f_\omega &= \int \frac{1}{r^2} \exp\left[ -\frac{r}{\zeta_\omega} - \frac{eH}{2}(x^2 + y^2) \right] d^3 r , \\
f'_\omega &= -\int \frac{x^2 + y^2}{2r^3} \exp\left[ -\frac{r}{\zeta_\omega} - \frac{eH}{2}(x^2 + y^2) \right] d^3 r ;
\end{aligned} \tag{19.58}$$

cf. the step leading from Eq. (19.44) to Eq. (19.45). If also the step leading from Eq. (19.45) to Eq. (19.46) is repeated, we obtain

$$f_\omega = \frac{4\pi}{eH} \int_0^\infty \arctan(k\zeta_\omega) \exp\left( -\frac{k^2}{2eH} \right) dk . \tag{19.59}$$

Further it is

$$f'_\omega = \frac{1}{2eH} \left( \frac{f_\omega}{\zeta_\omega} - 4\pi \right) . \tag{19.60}$$

The last equation is proved by noticing that Eq. (19.59) may be written as

$$f_\omega = \zeta_\omega l_\omega (H \zeta_\omega^2) , \tag{19.61}$$

where $l_\omega$ is a function of the argument written down. It further follows from Eq. (19.58) that it is

$$\frac{\partial f'_\omega}{\partial \zeta_\omega} = \frac{1}{\zeta_\omega^2 e} \frac{\partial f_\omega}{\partial H} = \frac{1}{2eH} \frac{\partial}{\partial \zeta_\omega} \left( \frac{f_\omega}{\zeta_\omega} \right) . \tag{19.62}$$

The last equality sign is a consequence of Eq. (19.61). Eq. (19.62) may, for $H$ fixed, easily be integrated since it is

$$f'_\omega(\zeta_\omega = 0) = 0 , \tag{19.63}$$

which follows from the second Eq. (19.58), and

$$\lim_{\zeta_\omega \to 0} \frac{f_\omega}{\zeta_\omega} = 4\pi , \tag{19.64}$$

which follows by an expansion of the righthand side of Eq. (19.59).

In connection with the anisotropic part of the differential scattering cross section (19.51), we further need

$$\oint \frac{\mathbf{v} \cdot \mathbf{v'}}{v^2} f_\omega(x, \mathbf{v'}) \, d\Omega' = -2eH f'_\omega \frac{(v_x - iv_y)}{v} \times \Delta(\mathbf{r}) \qquad (19.65)$$

and

$$\oint \frac{\mathbf{v} \cdot \mathbf{v'}}{v^2} f'_\omega(x, \mathbf{v'}) \, d\Omega' = -\frac{f'_\omega}{\zeta_\omega} \frac{(v_x - iv_y)}{v} \times \Delta(\mathbf{r}). \qquad (19.66)$$

Also these relations are obtained with the help of the calculus of residues.

Inserting now everything into Eq. (19.52), we obtain a linear relation between $f_\omega(x, \mathbf{v})$ and $f'_\omega(x, \mathbf{v})$. Since these two functions are linearly independent with respect to their $\mathbf{v}$ dependence, we derive the system of linear equations,

$$a_\omega \left( 1 - \frac{f_\omega}{4\pi l} \right) - b_\omega \frac{f'_\omega}{4\pi l} = \frac{1}{(2\pi)^2 \, v},$$

$$a_\omega \frac{2eHc f'_\omega}{4\pi l} + b_\omega \left( 1 + \frac{c f'_\omega}{4\pi l \zeta_\omega} \right) = 0, \qquad (19.67)$$

for the coefficients, $a_\omega$ and $b_\omega$. It is solved by

$$a_\omega = \frac{1}{(2\pi)^2 \, v} \frac{1 + \dfrac{c f'_\omega}{4\pi l \zeta_\omega}}{1 - \dfrac{f_\omega}{4\pi l} + \dfrac{c f'_\omega}{4\pi l \xi_\omega}}, \qquad (19.68)$$

whereas $b_\omega$ is not needed.

The eigenvalue $\kappa_\omega$ [Eq. (19.19)] is now calculated from

$$\oint \Delta_\omega(\mathbf{r}, \mathbf{v}) \, d\Omega = \frac{\kappa_\omega}{\pi v} \Delta(\mathbf{r}); \qquad (19.69)$$

cf. Eq. (11.24). Eq. (19.57) leads to

$$\kappa_\omega = \pi v (a_\omega f_\omega + b_\omega f'_\omega) = l(a_\omega (2\pi)^2 \, v - 1), \qquad (19.70)$$

where the second equality sign follows from the first Eq. (19.67). Insertion of Eq. (19.68) gives the result

$$\kappa_\omega = \frac{f_\omega + c \dfrac{f'_\omega}{l}}{4\pi - \dfrac{f_\omega}{l} + c \dfrac{f'_\omega}{l \zeta_\omega}} \qquad (19.71)$$

For isotropic scattering ($c = 0$), the righthand side reduces to an expression already discussed, though not explicitly written down, above.

## 20. The Surface Normal Critical Field $H_{c\perp}$

Let a slice of a homogeneous metal be bounded by two plane parallel surfaces and be infinitely extended in the other directions. As a special limiting case, we may also imagine a semi-infinite space bounded by a plane surface and filled with a homogeneous metal. If a homogeneous magnetic field is applied in the direction normal to the surfaces (or the single surface), the critical field for a second kind transition in this geometry is called $H_{c\perp}$.

### a) Ginzburg-Landau and Diffusion Approximation

We define the surface normal as the $z$ direction and use the gauge (19.1) for the vector potential. The GL equation (19.2) has, on both boundaries, to be supplemented by the standard boundary condition

$$\frac{\partial \Delta(\mathbf{r})}{\partial z} = 0, \qquad (20.1)$$

which follows from Eq. (15.1). It is immediately seen that the solution (19.10) is applicable also to the present case. Therefore, it is

$$H_{c\perp}(T) = H_{c2}(T), \qquad (20.2)$$

whenever the GL approximation or the diffusion approximation is valid. We shall see in Section 20b that $H_{c\perp}$ and $H_{c2}$ are equal beyond the validity of these approximations provided that the boundaries reflect the electrons specularly. For non-specular reflexion and for conditions somewhat outside the regions of validity of the GL and diffusion approximation, correction terms to Eq. (20.2) shall be derived in Section 20c.

There is still the $k$ degeneracy connected with the pair potential (19.10). It is again, as in the $H_{c2}$ case (Section 19a), possible to form linear combinations of the form (19.15). Therefore, the Abrikosov pattern of currents and magnetic field appears also in the present problem.

The GL or diffusion approximations are, on the one hand, valid only if the temperature is sufficiently close to $T_c$ (GL approximation) or if the electronic mean free path is sufficiently short (diffusion approximation). On the other hand, the linear dimensions of the sample have to be sufficiently big. In the present problem this means that the thickness of the metallic layer must be big as compared to, roughly, $\sqrt{\xi_0 \zeta_0}$ (GL approximation) or to the electronic mean free path (diffusion approximation). The correction terms to Eq. (20.2), which will be derived in Section 20c, clearly indicate the various requirements for the validity of the GL and diffusion approximation.

*b) General Theorems*

We want to show [39] that Eq. (20.1) is quite generally true provided that the boundaries reflect electrons in a specular manner. This theorem seems of some interest though specular semi-microscopic boundary conditions are of a somewhat academic character. Actually, it shall be shown that also the pair potential (19.10) is still a correct solution to the problem.

We start from the $H_{c2}$ problem in an infinitely extended homogeneous metal (Section 19). The pair potential (19.10) is $z$ independent. Obviously also the distribution function $\Delta_\omega(\mathbf{r}, \mathbf{v})$ obeying

$$
\left( 2|\omega| + v_x \frac{\partial}{\partial x} + v_y \left( \frac{\partial}{\partial y} + 2ieHx \right) + v_z \frac{\partial}{\partial z} + nv\sigma \right) \Delta_\omega(\mathbf{r}, \mathbf{v})
$$
$$
- nv \oint \frac{d\sigma(\theta)}{d\Omega} \Delta_\omega(\mathbf{r}, \mathbf{v}') \, d\Omega' = \frac{1}{(2\pi)^2} \Delta(\mathbf{r}) \tag{20.3}
$$

is then $z$ independent. Therefore, also the sign of $v_z$ is irrelevant. The unique solution to Eq. (20.3) satisfies

$$
\Delta_\omega(\mathbf{r}, v_x, v_y, v_z) = \Delta_\omega(\mathbf{r}, v_x, v_y, -v_z). \tag{20.4}
$$

This equation is identical with the boundary condition for specular reflexion by a plane surface, normal to the $z$ direction. The infinite material may, therefore, be cut into slices with specularly reflecting plane boundaries without modifying the distribution function $\Delta_\omega(\mathbf{r}, \mathbf{v})$. Application of Eq. (13.5) then closes the argument: The pair potential (19.10) satisfies the gap equation (10.1) also in the present problem, and Eq. (20.1) follows immediately. It should perhaps be emphasized that the critical field is equal to $H_{c2}$, only if the magnetic field is applied normal to the plane boundaries of the sample.

The second theorem [53] to be discussed states that Eq. (19.10) is modified in the form

$$
\Delta(\mathbf{r}) = \exp\left[ -eH \left( x + \frac{k}{2eH} \right)^2 + iky \right] g(z) \tag{20.5}
$$

for more general microscopic boundary conditions. Eq. (19.10) itself, which is valid in the GL and diffusion approximation and more generally for specular reflexion, obviously is a special case of Eq. (20.5). For the validity of Eq. (20.5) it is required that the reflexion kernel is the same in all points of a boundary (but it may be different on different boundaries) and that it is rotationally invariant about the surface normal.

The general arguing is very similar to the one presented in Section 19b. We choose a gauge in which $A_z$ vanishes and $A_x$ and $A_y$ do not depend upon $z$. We may in particular use the gauge (19.1) for which Eq. (20.5) was written down. If again a function $\tilde{K}_\omega(r, r')$ is introduced with the help of Eq. (19.21), it depends only upon the translationally and rotationally invariant quantity $(x - x')^2 + (y - y')^2$ and upon $z$, $z'$, and $H$. The integral on the righthand side of Eq. (19.21) is effectively extended in the $x - y$ plane along the straight line from $(x', y')$ to $(x, y)$. We then show that $\mathbb{K}_\omega$ commutes with the operator $(n \times \tilde{\mathbb{p}})^2$. Therefore, eigenfunctions to $\sum_\omega' \mathbb{K}_\omega$ may be chosen such that they are also eigenfunctions to $(n \times \tilde{\mathbb{p}})^2$. In the gauge (19.1), this means that $\Delta(r)$ satisfies

$$\left[ \frac{\partial^2}{\partial x^2} + \left( \frac{\partial}{\partial y} + 2ieHx \right)^2 \right] \Delta(r) = -2eH\, \Delta(r), \tag{20.6}$$

where the number $n$ in the sense of Eq. (19.6) is again chosen equal to zero. Since $K_\omega(r, r')$, according to the arguments given above, depends upon $y$ and $y'$ only through the combination $y - y'$, the $y$ dependence of $\Delta(r)$ may be chosen as a factor $\exp(iky)$. This remark essentially finishes the proof of Eq. (20.5). It may again be shown that the actual choice of $k$ is irrelevant for the problem.

The main further task is the simultaneous determination of the critical field, $H_{c\perp}(T)$, and of the function $g(z)$.

### c) Going beyond the Approximations

If we go, in the absence of impurities, slightly beyond the GL approximation, the GL equation (13.1) or (19.2) is still valid. If Eq. (20.5) is inserted into Eq. (19.2), the latter becomes a differential equation for the function $g(z)$,

$$\frac{d^2 g(z)}{dz^2} = (2eH - \lambda(T))\, g(z). \tag{20.7}$$

For a properly selected $g(z)$, the magnetic field $H$ is equal to the critical field $H_{c\perp}(T)$, whereas $\lambda(T)$ is related to $H_{c2}(T)$ by Eq. (19.8). The boundary condition is in principle given by Eq. (15.32). If only the first term on the righthand side of that equation is retained, it reads

$$\frac{dg(z)}{dz} = -p\, \frac{45\zeta(4)}{224\zeta(3)}\, 2eH\, \xi_0(T_c)\, g(z) \tag{20.8}$$

on the surface with the smaller value of $z$. The sign on the righthand side is the opposite on the other surface. Eq. (20.6) was used in the derivation. The parameter $p$ either is the diffuseness in the sense of Eq. (10.24), or it is more generally defined by Eq. (15.42). In particular, it is $p = 0$ for specular reflexion so that the boundary condition reduces to Eq. (20.1), and Eq. (20.2) is valid. The parameter $p$ can be chosen different on the two surfaces but this shall not be done.

The highest magnetic field is obtained if Eq. (20.7) is solved by hyperbolic functions. For a film or layer of thickness $a$, the combination of Eqs. (20.7) and (20.8) leads to an implicit equation,

$$\sqrt{2eH - \lambda(T)} \tanh \sqrt{2eH - \lambda(T)} \frac{a}{2} = p \frac{45\zeta(4)}{224\zeta(3)} 2eH\xi_0(T_c), \quad (20.9)$$

for the field $H = H_{c\perp}(T)$. Since $H_{c\perp}$ becomes equal to $H_{c2}$ for $p = 0$, we may replace $H$ on the righthand side by $H_{c2}$. Simple results are only obtained in limiting cases. For

$$\sqrt{2eH - \lambda} \frac{a}{2} \ll 1, \quad (20.10)$$

it is

$$\frac{H_{c\perp}(T)}{H_{c2}(T)} = 1 + p \frac{45\zeta(4)}{224\zeta(3)} \frac{2\xi_0(T_c)}{a} + \cdots, \quad (20.11)$$

whereas, for

$$\sqrt{2eH - \lambda} \frac{a}{2} \gg 1, \quad (20.12)$$

it is

$$\frac{H_{c\perp}(T)}{H_{c2}(T)} = 1 + \left( p \frac{45\zeta(4)}{224\zeta(3)} \right)^2 2eH_{c2}(T) \xi_0^2(T_c) + \cdots \quad (20.13)$$

or

$$\frac{H_{c\perp}(T)}{H_{c2}(T)} = 1 + \left( p \frac{45\zeta(4)}{224\zeta(3)} \right)^2 \frac{12}{7\zeta(3)} \frac{T_c - T}{T_c} + \cdots. \quad (20.14)$$

The latter relation was obtained from Eq. (20.13) with the help of Eqs. (19.8) and (13.13).

Eq. (20.8) defines a characteristic length,

$$f(T) = \frac{1}{p} \frac{224\zeta(3)}{45\zeta(4)} \frac{1}{2eH_{c2}(T) \xi_0^2(T_c)} \xi_0(T_c), \quad (20.15)$$

which is strongly temperature dependent as a consequence of Eqs. (19.8) and (13.13),

$$f(T) = \frac{1}{p} \left( \frac{14\zeta(3)}{3} \right)^2 \frac{2}{15\zeta(4)} \frac{T_c}{T_c - T} \xi_0(T_c). \quad (20.16)$$

It follows from Eq. (20.11) that condition (20.10) may also be written as

$$\frac{a}{2} \ll f, \quad (20.17)$$

whereas condition (20.12) may be written as

$$\frac{a}{2} \gg f. \quad (20.18)$$

The prototype of condition (20.18) is a semi-infinite sample filling all space with $z \geq 0$. In this case, the pair potential decays exponentially,

$$g(z) \propto \exp\left(-\frac{z}{f}\right) \tag{20.19}$$

behind the surface. In the opposite limit (20.17), the function $g(z)$ is practically constant in the sample.

The correction term in Eq. (20.14) arises from the fact that $T$ has to be very close to $T_c$ in order that the GL approximation is valid. The reason for the temperature independent correction term in Eq. (20.11) is that all linear dimensions of the sample (including its thickness) have to be big as compared to $\xi_0$, the thickness of the surface layer discussed in Sec. 15a. That the correction terms vanish for specular reflexion ($p = 0$) is a consequence of the theorem that Eq. (20.2) is quite generally valid in this case.

The analysis reported so far is valid only in the absence of impurities. But the method of Section 17 permits to treat metals containing impurities as well, provided that their concentration is homogeneous and the differential scattering cross section is (or is assumed to be) isotropic. Eq. (17.39) is the result of a perturbation treatment. $\Delta(r)$, on the righthand side, may be taken from Eq. (19.42). The integrals in numerator and denominator both diverge if they are taken literally. Since the integrands do not depend upon $y$, the $y$ integrations shall be discarded. The integration is, therefore, only extended along the $x$ axis in the numerator. It is extended along the $x$ axis and over $z$ inside the sample, in the denominator. The elementary result is

$$\frac{\oint |n \times \tilde{\partial} \Delta(r)|^2 |d^2 r|}{\int |\Delta(r)|^2 d^3 r} = \frac{4eH}{a}. \tag{20.20}$$

Further it is

$$\frac{d\lambda}{dH} = 2e \tag{20.21}$$

as a consequence of Eq. (19.8).

In this way, we arrive at the result

$$\delta H = p F_1\left(\frac{\xi_0}{l}\right) H_{c2}(T) \frac{2\xi_0(T)}{a} \tag{20.22}$$

or

$$\frac{H_{c\perp}(T)}{H_{c2}(T)} = 1 + p F_1\left(\frac{\xi_0}{l}\right) \frac{2\xi_0(T)}{a} + \cdots . \tag{20.23}$$

12*

The function $F_1(\xi_0/l)$ was defined in Eq. (17.41). It is shown in Fig. 3 on p. 156. Because of Eq. (17.42), Eq. (20.11) is rederived in the absence of impurities. In the dirty limit [Eq. (14.15)], Eq. (17.43) leads to the result

$$\frac{H_{c\perp}(T)}{H_{c2}(T)} = 1 + p\,\frac{3l}{8a} + \cdots, \tag{20.24}$$

which should be valid at all temperatures $T \le T_c$. The temperature independent correction term in Eq. (20.24) reflects the fact, that the linear dimensions of the sample have to be big as compared to the electronic mean free path in order that the diffusion approximation is valid. The treatment given here, evidently corresponds to the limiting case (20.17) where, however, the characteristic length, $f$, still remains to be determined.

More general situations, in particular the opposite limit (20.18), could easily be treated if the local boundary condition (17.60) were regarded as justified. It would be seen that Eq. (20.15) is now replaced by

$$f = \left[ pF_1\left(\frac{\xi_0}{l}\right) 2eH_{c2}\,\xi_0^2(T) \right]^{-1} \xi_0(T). \tag{20.25}$$

For $T$ close to $T_c$, we would obtain

$$f = \frac{1}{p}\,\frac{7\zeta(3)}{12}\,\frac{\chi\left(\dfrac{\xi_0}{l}\right)}{F_1\left(\dfrac{\xi_0}{l}\right)}\,\frac{T_c - T}{T_c}\,\xi_0(T_c) \tag{20.26}$$

with the help of Eqs. (19.8) and (13.26). In the dirty limit when local boundary conditions are certainly justified, Eq. (20.25) becomes

$$f = \frac{1}{p}\,\frac{16}{3}\,\frac{1}{2eH_{c2}(T)\,l}. \tag{20.27}$$

In this case, it is

$$f = \frac{1}{p}\,\frac{32\gamma}{9}\,\xi_0(T_c) \tag{20.28}$$

for $T = 0$, as follows from Eqs. (19.8), (14.34), and (14.40). Therefore, the length $f$ is big as compared to $l$, the thickness of the surface layer in the sense of Section 16, even for $T = 0$.

The correction terms are now easily obtained if Eq. (20.19) is inserted into Eq. (20.7). It is

$$\frac{H_{c\perp}(T)}{H_{c2}(T)} = 1 + \frac{1}{2eH_{c2}(T)\,f^2} + \cdots \tag{20.29}$$

or, with application of Eq. (20.25),

$$\frac{H_{c\perp}(T)}{H_{c2}(T)} = 1 + \left[ p F_1 \left( \frac{\xi_0}{l} \right) \right]^2 2 e H_{c2} \xi_0^2(T) + \cdots . \tag{20.30}$$

From this equation we obtain

$$\frac{H_{c\perp}(T)}{H_{c2}(T)} = 1 + \frac{\left[ p F_1 \left( \dfrac{\xi_0}{l} \right) \right]^2}{\chi \left( \dfrac{\xi_0}{l} \right)} \frac{12}{7 \zeta(3)} \frac{T_c - T}{T_c} + \cdots \tag{20.31}$$

for $T$ close to $T_c$ (GL approximation) and

$$\frac{H_{c\perp}(T)}{H_{c2}(T)} = 1 + p^2 \left( \frac{3}{16} \right)^2 2 e H_{c2}(T) \, l^2 + \cdots \tag{20.32}$$

in the dirty limit (diffusion approximation). In the latter case the correction term is of the order of $l/\xi_0(T_c)$ for $T = 0$. It is a small quantity because of Eq. (14.15).

A, perhaps, more reliable derivation of the same results is obtained by a slight modification of the method of Sections 17c. We incorporate the boundary condition

$$n \cdot \tilde{\partial} \Delta(r) = \frac{1}{f} \Delta(r) \tag{20.33}$$

into the unperturbed kernel $\mathbb{K}_\omega^{(\sigma)}$, so that Eq. (17.23) is replaced by

$$(\Delta, \mathbb{K}_\omega^{(\sigma)} \Delta) = \frac{2 \pi N \zeta_\omega}{v} \left\{ \int \left[ |\Delta(r)|^2 - \frac{\zeta_\omega^2}{3} |\tilde{\partial} \Delta(r)|^2 \right] d^3 r + \frac{\zeta_\omega^2}{3} \oint \frac{1}{f} |\Delta(r)|^2 |d^2 r| \right\}, \tag{20.34}$$

where $f$ remains to be determined in a later step of the calculation. The surface term in Eq. (20.34) has again to be subtracted in $(\Delta, \delta \mathbb{K}_\omega^{(\sigma)} \Delta)$ [Eq. (17.37)] so that Eq. (17.39) is replaced by

$$\delta H = - \frac{1}{d\lambda/dH} \frac{\oint \dfrac{1}{f} |\Delta(r)|^2 |d^2 r|}{\int |\Delta(r)|^2 d^3 r} + \cdots , \tag{20.35}$$

where the dots stand for the righthand side of Eq. (17.39). Since $H_{c\perp}(T)$ is already modified by the boundary condition (20.33), we obtain

$$2 e (H_{c\perp} - H_{c2}) = - \frac{1}{f^2} + 2 p \, 2 e H_{c2} F_1 \left( \frac{\xi_0}{l} \right) \frac{\xi_0}{f} . \tag{20.36}$$

$H_{c\perp}$ is now maximized with respect to the parameter $f$. This maximization requirement finds its systematic justification on the basis of the variational principle connected with the functional (17.30). Eq. (20.25) gives the optimal value of $f$. The further evaluation is done in essentially the same way as above. – Another derivation of Eq. (20.30) is given in Ref. [53].

That the correction term $C(T)$ in Eq. (17.33) was neglected in the calculations, still requires a justification. Strictly speaking, a correction term due to $C(T)$ enters into the expressions both for $H_{c2}(T)$ and for $H_{c\perp}(T)$. These correction terms, however, cancel in the present approximation if the ratio, $H_{c\perp}(T)/H_{c2}(T)$, is formed.

Finally it might be asked whether only $H_{c\perp}$ or also $H_{c2}$ may be measured in this geometry on a sufficiently thick sample [with Eq. (20.18) well satisfied]. The answer is obviously the following. If we start with a magnetic field strong enough to suppress superconductivity entirely, and lower it continuously, superconductivity will first appear when the strength of the magnetic field becomes equal to $H_{c\perp}(T)$. When the field is further lowered, the theory linear in the pair potential, is no longer applicable. The Gibbs free energy in the sense of Section 2 will become smaller in the superconducting state than in the normal state, the difference between the two values being proportional to the surface of the sample.

A new type of solution to Eqs. (20.7) and (20.8) (the latter generalized to the presence of impurities, if necessary) turns up at a magnetic field slightly lower than $H_{c2}$. More exactly, this field is given by

$$H = H_{c2} - \frac{\pi^2}{2\,e\,a^2}, \tag{20.37}$$

if a further, $f$ dependent, correction term is neglected. The $a$ dependent correction term in Eq. (20.37) is negligeable provided that $a$ is big as compared to $\xi(T)$. The new solution is a trigonometric function. It penetrates into the interior of the sample. If the magnetic field is further lowered, the Gibbs free energy difference between normal and superconducting state, corresponding to the new solution, is proportional to the volume of the sample. The Gibbs free energy corresponding to the new solution will finally, and for a sufficiently thick sample still close to $H_{c2}$, become smaller than that corresponding to the original solution. Therefore, another phase transition, not strictly of the second kind, occurs at a magnetic field slightly below $H_{c2}$. Similar considerations may be made in connection with the $H_{c3}$ problem to be studied in the next section.

Ref. [53] contains a theoretical analysis of the $H_{c\perp}$ problem which goes beyond the one given in this section.

## 21. The Surface Parallel Critical Field $H_{c3}$

If a semi-infinite sample is bounded by a plane surface and if a homogeneous magnetic field is applied parallel to the surface, the critical field for a second kind transition is called $H_{c3}(T)$.

## a) Ginzburg-Landau and Diffusion Approximation

The field $H_{c3}$ was discovered theoretically in this approximation by *Saint-James* and *de Gennes* [54]. They derived the simple result

$$H_{c3}(T) = 1.695\, H_{c2}(T)\,. \tag{21.1}$$

Our first aim is to reproduce this result.

The Cartesian coordinate system shall be chosen such that the homogeneous metal fills the semi-infinite space $x \geq 0$ and that the homogeneous magnetic field is parallel to the $z$ direction. If the gauge is again chosen according to Eq. (19.1), the GL equation is still given by Eq. (19.2). Again, the ansatz (19.3) may be made where $f(x)$ satisfies Eq. (19.5). The function $f(x)$ has to be bounded for $x \to +\infty$. The boundary condition at $x = 0$ is given by

$$\frac{d\,f(0)}{dx} = 0\,, \tag{21.2}$$

as follows from Eq. (15.1). There is obviously no longer a degeneracy with respect to $k$ [Eq. (19.3)], but this parameter has to be chosen in such a way that the magnetic field is as big as possible. The other parameter, $\mu$, has again to be put equal to zero.

If the parameter $k$ is, for a fixed number of zeros of the function $f(x)$ in the interval $0 < x < +\infty$, exhibited, we have to postulate

$$\frac{\partial f_k(0)}{\partial x} = 0\,, \qquad \frac{\partial^2 f_k(0)}{\partial k\, \partial x} = 0 \tag{21.3}$$

at the surface, $x = 0$. The first equation, which is identical with Eq. (21.2), is an implicit relation between $H$ and $k$. For the maximal value of $H$ ($dH/dk = 0$), also the second equation (21.3) is true. Since it follows from Eq. (19.5) that a solution bounded at $x \to +\infty$ depends upon $x$ and $k$ only through the combination $x + k/2eH$, Eq. (21.3) may also be written as

$$\frac{d\,f(0)}{dx} = 0\,, \qquad \frac{d^2 f(0)}{dx^2} = 0\,. \tag{21.4}$$

*Saint-James* and *de Gennes* showed that the correct solution is the one without zeros and that it is

$$2eH_{c3}(T) = 1.695\, \lambda(T)\,. \tag{21.5}$$

Eq. (21.1) follows immediately, if also Eq. (19.8) is taken into account.

A combination of Eq. (19.5), with $\mu$ put equal to zero, and of the second Eq. (21.4) leads to the result

$$k^2 = \lambda \tag{21.6}$$

for the correct solution. Actually, it turns out that $k$ is a negative quantity, $k = -\sqrt{\lambda}$. If the function $f(x)$ is chosen non-negative, it decreases monotonically in the interval $0 \leq x < +\infty$. A rough estimate of the thickness of the surface sheath is obviously given by

$$\frac{|k|}{2eH} = \frac{\sqrt{\lambda}}{2eH} = \frac{\xi(T)}{1.695}, \tag{21.7}$$

where $\lambda$ was introduced from Eq. (21.6) and $\xi(T)$ from Eq. (13.2). The number, 1.695, is of course irrelevant for the estimate.

The electric current density is again given by Eq. (19.11). The pair potential, $\Delta(r)$, has to be taken from Eq. (19.3) with $\mu$ put equal to zero. The $x$ component of the current density vanishes since $f(x)$ may be chosen as a real function. The $z$ component vanishes for $\mu = 0$. Therefore, the electric current flows parallel to the surface and orthogonal to the direction of the magnetic field. It is

$$i_y(r) \propto -2(k + 2eHx)|f(x)|^2 . \tag{21.8}$$

The current density, therefore, depends only upon the distance, $x$, from the surface. The current flows in the positive $y$ direction for

$$0 \leq x < \frac{|k|}{2eH} = \frac{\xi(T)}{1.695} \tag{21.9}$$

[cf. Eq. (21.7)]. It flows in the opposite direction for bigger values of $x$. The net current vanishes,

$$\int_0^\infty (k + 2eHx)|f(x)|^2 \, dx = 0 . \tag{21.10}$$

This result follows, for real $f(x)$, from

$$\int_0^\infty \frac{d}{dx} [((k + 2eHx)^2 - \lambda) f^2(x)] \, dx$$

$$= 4eH \int_0^\infty (k + 2eHx) f^2(x) \, dx + 2 \int_0^\infty [(k + 2eHx)^2 - \lambda] f(x) \frac{df(x)}{dx} \, dx . \tag{21.11}$$

The integral on the lefthand side vanishes because of Eq. (21.6), and since $f(x)$ goes sufficiently strongly to zero for $x \to +\infty$. The second integral on the righthand side vanishes if Eq. (19.5), with $\mu = 0$, is inserted and Eq. (21.2) is applied.

## b) Going beyond the Approximations

For a position independent reflexion kernel, the kernel $K_\omega(r, r')$ depends only upon $x$, $x'$, $y - y'$, and $z - z'$ in the gauge (19.1). This assertion follows from the decomposition (19.21), with the exponential given by Eq. (19.34). Therefore, the ansatz (19.3) may also be made in the general case.

At present we only want to go somewhat beyond the approximations made in Section 21a. Eq. (19.5) is still valid. The boundary condition (15.32), which is valid for a clean metal, becomes a boundary condition for the function $f(x)$,

$$\frac{d f(0)}{dx} = - \left[ p \frac{45\,\zeta(4)}{224\,\zeta(3)} \xi_0(T)\,k^2 + \left(1 + \frac{p'}{2}\right) \frac{31\,\zeta(5)}{70\,\zeta(3)} \xi_0^2(T)2eHk \right] f(0) + \cdots$$

$$(21.12)$$

This relation replaces Eq. (21.2).

It is convenient for the general analysis to introduce dimensionless variables[*],

$$\xi := \sqrt{2eH}\,x + \frac{k}{\sqrt{2eH}}, \qquad \varepsilon := \frac{\lambda}{2eH}, \qquad (21.13)$$

so that the surface $(x = 0)$ is now situated at

$$\xi = \frac{k}{\sqrt{2eH}}. \qquad (21.14)$$

Further we put

$$f(x) =: u(\varepsilon, \xi), \qquad (21.15)$$

where also the $\varepsilon$ dependence is exhibited. $u(\varepsilon, \xi)$ obeys the differential equation

$$\left(\frac{\partial^2}{\partial\xi^2} - \xi^2\right) u(\varepsilon, \xi) = -\varepsilon u(\varepsilon, \xi) \qquad (21.16)$$

as a consequence of Eq. (19.5) with $\mu$ put equal to zero. It is required to be bounded, and actually vanishes, for $\xi \to +\infty$.

In the $H_{c2}$ problem (Section 19), $u(\varepsilon, \xi)$ has to be bounded also for $x \to -\infty$. In this case, it is

$$\varepsilon = 1. \qquad (21.17)$$

The $H_{c3}$ treatment of Section 21a is obtained, if it is required

$$\frac{\partial u(\varepsilon, \xi)}{\partial\xi} = 0 = \frac{\partial^2 u(\varepsilon, \xi)}{\partial\xi^2} \qquad (21.18)$$

at the boundary, the position of which in terms of $\xi$ still remains to be determined. Eq. (21.18) is the translation of Eq. (21.4) into the new variables. For the nodeless function $u(\varepsilon, \xi)$, it is

$$\varepsilon = \varepsilon^{(0)} = \frac{1}{1.695}, \qquad (21.19)$$

---

[*] The reader is warned not to confuse the dimensionless variable $\xi$ with lengthes like $\xi(T)$, $\xi_0(T)$, and $\xi_\omega$.

which corresponds to Eq. (21.5). The surface is situated at

$$\xi = \xi^{(0)} = -\varepsilon^{(0)\frac{1}{4}} .  \tag{21.20}$$

This relation essentially corresponds to Eq. (21.6).

Boundary condition (21.12) is translated into

$$\frac{\partial u(\varepsilon, \xi)}{\partial \xi} = -\left[ p \frac{45\zeta(4)}{224\zeta(3)} \eta \xi^2 + \left( 1 + \frac{p'}{2} \right) \frac{31\zeta(5)}{70\zeta(3)} \eta^2 \xi \right] u(\varepsilon, \xi) + \cdots ,  \tag{21.21}$$

where the notation

$$\eta = \xi_0(T) \sqrt{2eH}  \tag{21.22}$$

was introduced. In the next step, only the first term on the righthand side of Eq. (21.21) shall be retained for $p \neq 0$. Only for specular reflexion, $p = p' = 0$, we take the second term into account. We also note the translation of Eq. (13.58),

$$\eta^2 \varepsilon = \frac{12}{7\zeta(3)} \frac{T_c - T}{T_c} + \left[ \frac{6}{7\zeta(3)} + \frac{31\zeta(5)}{10} \left( \frac{6}{7\zeta(3)} \right)^3 \right] \left( \frac{T_c - T}{T_c} \right)^2$$
$$+ \frac{93\zeta(5)}{140\zeta(3)} \eta^4 + \cdots .  \tag{21.23}$$

It will turn out that only the first term on the righthand side, which corresponds to Eq. (13.13), has to be retained for $p \neq 0$.

We expand the lefthand side of Eq. (21.21) about $\varepsilon = \varepsilon^{(0)}, \xi = \xi^{(0)}$ and make use of Eq. (21.18),

$$\frac{\partial u(\varepsilon, \xi)}{\partial \xi} = (\varepsilon - \varepsilon^{(0)}) \frac{\partial^2 u(\varepsilon, \xi)}{\partial \varepsilon \partial \xi} + \cdots .  \tag{21.24}$$

In this way, we obtain

$$\varepsilon - \varepsilon^{(0)} = \begin{cases} -p \dfrac{45\zeta(4)}{224\zeta(3)} C \eta \xi^{(0)2} + \cdots , \\[4mm] -\dfrac{31\zeta(5)}{70\zeta(3)} C \eta^2 \xi^{(0)} + \cdots , \end{cases}  \tag{21.25}$$

where the first line is true for non-specular reflexion $(p \neq 0)$ and the second one for specular reflexion. The symbol $C$ is shorthand for*

$$C = \left( \frac{u}{\dfrac{\partial^2 u}{\partial \varepsilon \partial \xi}} \right)_{\varepsilon^{(0)}, \xi^{(0)}} = 0.762 .  \tag{21.26}$$

---

* The numerical value was calculated by H. Schultens.

For $p \neq 0$, Eq. (21.23) may be used in the form

$$\eta^2 \varepsilon^{(0)} = \frac{12}{7\zeta(3)} \frac{T_c - T}{T_c} - \eta^2 (\varepsilon - \varepsilon^{(0)}) + \cdots . \tag{21.27}$$

If the first line of Eq. (21.25) is inserted, the resulting equation recursively solved for $\eta$, and the result combined with the corresponding $H_{c2}$ result [Eq. (21.17)], we obtain [55]

$$\frac{H_{c3}(T)}{1.695 H_{c2}(T)} = 1 + p \frac{45\zeta(4)}{112\zeta(3)} \left(\frac{3}{7\zeta(3)}\right)^{\frac{1}{4}} \frac{C}{\varepsilon^{(0)\frac{1}{2}}} \left(\frac{T_c - T}{T_c}\right)^{\frac{1}{4}}$$

$$= 1 + 0.214 p \left(\frac{T_c - T}{T_c}\right)^{\frac{1}{4}} + \cdots . \tag{21.28}$$

For $p = 0$, the analogue to Eq. (21.27) has to be used with all terms on the righthand side of Eq. (21.23) retained. If the same terms are also taken into account in the calculation of $H_{c2}$, the result is [55]

$$\frac{H_{c3}(T)}{1.695 H_{c2}(T)} = 1 + \frac{279\zeta(5)}{245\zeta^2(3)} \left(\frac{1 - \varepsilon^{(0)2}}{\varepsilon^{(0)2}} - \frac{2C}{3\varepsilon^{(0)\frac{1}{2}}}\right) \frac{T_c - T}{T_c} + \cdots$$

$$= 1 + 0.614 \frac{T_c - T}{T_c} + \cdots . \tag{21.29}$$

Eq. (21.29) is identical with a result obtained earlier by *Ebneth* and *Tewordt* [56]. An expansion, which combines Eqs. (21.28) and (21.29), will be given below in Eq. (21.45).

The correction terms are due to the fact, that the GL equation with boundary condition (15.1) is valid only for $T \to T_c$. The correction term is one order of magnitude smaller for $p = 0$. This observation is easily understood, since only the first term on the righthand side of Eq. (15.2) satisfies arbitrary semi-microscopic boundary conditions but the first two terms satisfy the boundary condition of specular reflexion. The correction term $\Delta_{\omega 1}(r, v)$, in the sense of Eq. (15.6), is one order of magnitude smaller for specular reflexion. Comparing Eq. (21.28) for $H_{c3}$ with Eq. (20.14) for $H_{c\perp}$ we see that the powers of $p$ and of $(T_c - T)/T_c$ are different in the two cases. But also this difference finds its interpretation, if it is assumed that the correction term in any case is of the order of $p$ times $\xi_0(T_c)/$(characteristic length). The characteristic length is the depth of the surface sheath, which is approximated by $\xi(T)$ for $H_{c3}$ and by $f$ [Eq. (20.16)] for $H_{c\perp}$.

The calculation may, for a metal containing (isotropically scattering) impurities, be based upon the method of Section 17. We have to insert

$$\frac{d\lambda}{dH} = \frac{2e}{1.695} \tag{21.30}$$

into Eqs. (17.39) and (17.47). Contributions from $C(T)$ in Eq. (17.33) cancel in the final result in the same way as the term proportional to $(T_c - T)^2/T_c^2$ in Eq. (21.23) did not contribute to Eq. (21.29). Further it is

$$\frac{\int |\boldsymbol{n} \times \tilde{\partial}\varDelta(\boldsymbol{r})|^2 \, |\mathrm{d}^2 r|}{\int |\varDelta(\boldsymbol{r})|^2 \, \mathrm{d}^3 r} = (2eH)^{\frac{1}{2}} \, \xi^{(0)2} \, \frac{|u(\varepsilon^{(0)}, \xi^{(0)})|^2}{\int\limits_{\xi^{(0)}}^{\infty} |u(\varepsilon^{(0)}, \xi)|^2 \, \mathrm{d}\xi} \tag{21.31}$$

and

$$-\frac{2ie}{\int |\varDelta(\boldsymbol{r})|^2 \, \mathrm{d}^3 r} \int (\mathrm{d}^2 r \times \boldsymbol{H}) \cdot (3\varDelta^*(\boldsymbol{r}) \, \tilde{\partial}\varDelta(\boldsymbol{r}) + (\tilde{\partial}\varDelta(\boldsymbol{r}))^* \, \varDelta(\boldsymbol{r}))$$

$$\tag{21.32}$$

$$= 2(2eH) \, \xi^{(0)} \, \frac{|u(\varepsilon^{(0)}, \xi^{(0)})|^2}{\int\limits_{\xi^{(0)}}^{\infty} |u(\varepsilon^{(0)}, \xi)|^2 \, \mathrm{d}\xi} \, .$$

Finally, the quantity $C$ is introduced from Eq. (21.26) by noticing that it is

$$C = \frac{|u(\varepsilon^{(0)}, \xi^{(0)})|^2}{\int\limits_{\xi^{(0)}}^{\infty} |u(\varepsilon^{(0)}, \xi)|^2 \, \mathrm{d}\xi} \, . \tag{21.33}$$

The last equation is obtained by combining Eq. (21.16) and its derivative,

$$\left(\frac{\partial^2}{\partial\xi^2} - \xi^2 + \varepsilon\right) \frac{\partial u(\varepsilon, \xi)}{\partial\varepsilon} = -u(\varepsilon, \xi) \, , \tag{21.34}$$

in an appropriate manner. It follows

$$\frac{\partial}{\partial\xi} \left(\frac{\frac{\partial u}{\partial\varepsilon}}{u}\right)_{\varepsilon\xi} = \frac{1}{u^2(\varepsilon, \xi)} \int\limits_{\xi}^{\infty} u^2(\varepsilon, \xi') \, \mathrm{d}\xi' \, . \tag{21.35}$$

The lefthand side reduces to an expression appearing in Eq. (21.26), for

$$\frac{\partial u(\varepsilon, \xi)}{\partial\xi} = 0 \, . \tag{21.36}$$

If now also Eqs. (21.5), (19.8), and (13.26) are inserted into Eqs. (17.39) and (17.47), we obtain

$$\frac{H_{c3}(T)}{1.695 H_{c2}(T)} = 1 + p \, \frac{F_1\left(\dfrac{\xi_0}{l}\right)}{\chi^{\frac{1}{2}}\left(\dfrac{\xi_0}{l}\right)} \left(\frac{12}{7\zeta(3)}\right)^{\frac{1}{2}} \frac{C}{\varepsilon^{(0)\frac{1}{2}}} \left(\frac{T_c - T}{T_c}\right)^{\frac{1}{2}} + \cdots \tag{21.37}$$

for $p \neq 0$, and

$$\frac{H_{c3}(T)}{1.695 H_{c2}(T)} = 1 + \frac{F_2\left(\frac{\xi_0}{l}\right)}{\chi\left(\frac{\xi_0}{l}\right)} \frac{36}{7\zeta(3)} K \frac{T_c - T}{T_c} + \cdots, \quad (21.38)$$

with

$$K = \frac{1 - \varepsilon^{(0)2}}{\varepsilon^{(0)2}} - \frac{2C}{3\varepsilon^{(0)\frac{1}{2}}} = 0.752, \quad (21.39)$$

for $p = 0$. The functions $F_1(\varrho)$ and $F_2(\varrho)$ where given in Eqs. (17.41) and (17.49), respectively. The function $\chi(\varrho)$ was defined in Eq. (13.28). The functions were shown in Figs. 3, 4, and 1 on ps. 156, 158, and 109. The length $\xi_0$ has at all places to be inserted as $\xi_0(T_c)$. Eqs. (21.37) and (21.38) reduce to Eqs. (21.28) and (21.29), respectively, in the clean limit because of Eqs. (17.42), (17.50), and $\chi(0) = 1$. In the dirty limit, we obtain

$$\frac{H_{c3}(T)}{1.695 H_{c2}(T)} = 1 + p \frac{3\sqrt{3}}{8\pi} \frac{C}{\varepsilon^{(0)\frac{1}{2}}} \left(\frac{l}{\xi_0(T_c)}\right)^{\frac{1}{2}} \left(\frac{T_c - T}{T_c}\right)^{\frac{1}{2}} + \cdots$$

$$= 1 + 0.205 p \left(\frac{l}{\xi_0(T_c)}\right)^{\frac{1}{2}} \left(\frac{T_c - T}{T_c}\right)^{\frac{1}{2}} + \cdots \qquad (21.40)$$

and

$$\frac{H_{c3}(T)}{1.695 H_{c2}(T)} = 1 + \frac{36}{5\pi^2} K \frac{l}{\xi_0(T_c)} \frac{T_c - T}{T_c} + \cdots$$

$$= 1 + 0.549 \frac{l}{\xi_0(T_c)} \frac{T_c - T}{T_c} + \cdots \qquad (21.41)$$

$(p = 0)$ from Eqs. (17.43), (17.51), and (13.29). The correction terms are much smaller in the dirty limit than in the clean limit. Identical results were, for the case of specular reflexion, obtained earlier by *Ebneth* and *Tewordt* [56].

In the dirty limit in the sense of Eq. (14.15), the diffusion approximation is valid at all temperatures $T \leq T_c$. Eq. (13.26) may now, of course, not be used. The result of the corresponding calculation is

$$\frac{H_{c3}(T)}{1.695 H_{c2}(T)} = 1 + p \frac{3}{16} \frac{C}{\varepsilon^{(0)\frac{1}{2}}} [2 e H_{c2}(T) l^2]^{\frac{1}{2}} + \cdots$$

$$= 1 + 0.186 p [2 e H_{c2}(T) l^2]^{\frac{1}{2}} + \cdots \qquad (21.42)$$

$(p \neq 0)$ and

$$\frac{H_{c3}(T)}{1.695 H_{c2}(T)} = 1 + \frac{3K}{5} 2 e H_{c2}(T) l^2 + \cdots$$

$$= 1 + 0.451 [2 e H_{c2}(T) l^2] + \cdots \qquad (21.43)$$

($p = 0$). For the discussion of the magnitude of the correction terms it should be kept in mind that $2eH_{c2}(T)\,l\xi_0(T_c)$ is of order unity for $T$ close to zero.

Eq. (21.28) is not very useful for a comparison with experimental results. The numerical factor of the first correction term is rather small as compared to the next term, the order of magnitude of which may be inferred from the correction term in Eq. (21.29). It is, indeed, possible to go one step further in the series expansion [57]. The boundary condition is now given by the complete Eq. (21.21). In order to maximize the magnetic field, we also have to postulate

$$\frac{\partial^2 u(\varepsilon, \xi)}{\partial \xi^2} = -p\,\frac{45\,\zeta(4)}{224\,\zeta(3)}\,\eta\,\frac{\partial}{\partial \xi}\,(\xi^2\,u(\varepsilon, \xi)) - \cdots. \qquad (21.44)$$

This is a generalization of the second condition (21.18). A straightforward calculation, which needs not be reproduced here, leads to the result

$$\frac{H_{c3}(T)}{1.695\,H_{c2}(T)} = 1 + p\,\frac{45\,\zeta(4)}{112\,\zeta(3)}\left(\frac{3}{7\,\zeta(3)}\right)^{\frac{1}{2}}\frac{C}{\varepsilon^{(0)\frac{1}{2}}}\left(\frac{T_c - T}{T_c}\right)^{\frac{1}{2}}$$

$$+ \frac{3}{7\,\zeta(3)}\left[\frac{93\,\zeta(5)}{35\,\zeta(3)}\left(K - p'\,\frac{C}{3\,\varepsilon^{(0)\frac{1}{2}}}\right) + p^2\left(\frac{45\,\zeta(4)}{112\,\zeta(3)}\right)^2 K'\right]\frac{T_c - T}{T_c} + \cdots. \qquad (21.45)$$

$$= 1 + 0.214\,p\left(\frac{T_c - T}{T_c}\right)^{\frac{1}{2}} + (0.614 - 0.458\,p' + 0.126\,p^2)\frac{T_c - T}{T_c} + \cdots.$$

Here, it is

$$K' = \frac{3\,C^2}{2\,\varepsilon^{(0)}} + \frac{C}{\varepsilon^{(0)\frac{1}{2}}}\left(1 - \frac{C\,\varepsilon^{(0)\frac{1}{2}}}{2}\right)^2 + \frac{D\,C^3}{2} = 2.706 \qquad (21.46)$$

with

$$D = \frac{\partial}{\partial \varepsilon}\left(\frac{\dfrac{\partial^2 u}{\partial \varepsilon\,\partial \xi}}{u}\right)_{\varepsilon^{(0)}\,\xi^{(0)}} = 1.76\,. \qquad (21.47)$$

The numbers in Eqs. (21.26), (21.46), and (21.47) represent a comedy of errors. The value of $C$ was first given incorrectly in Ref. [55]. The correct value was subsequently communicated in Ref. [57], but the value of the newly introduced constant $D$ was here in error as pointed out by *Sarma* [58]. A reevaluation confirmed *Sarma*'s result. The graphical representation of Eq. (21.45), in Ref. [57], may still be used since the effective error due to the wrong value of $D$ is only small.

It is not difficult to conjecture a generalization to Eq. (21.45) with the help of Eqs. (17.42) and (17.50). The result would be

$$\frac{H_{c3}(T)}{1.695\,H_{c2}(T)} = 1 + p\,\frac{F_1\left(\frac{\xi_0}{l}\right)}{\chi^{\frac{1}{2}}\left(\frac{\xi_0}{l}\right)}\left(\frac{12}{7\zeta(3)}\right)^{\frac{1}{2}}\frac{C}{\varepsilon^{(0)\frac{1}{2}}}\left(\frac{T_c-T}{T_c}\right)^{\frac{1}{2}} \qquad (21.48)$$

$$+\frac{36}{7\zeta(3)}\,\frac{1}{\chi\left(\frac{\xi_0}{l}\right)}\left[F_2\left(\frac{\xi_0}{l}\right)\left(K-p'\,\frac{C}{3\,\varepsilon^{(0)\frac{1}{2}}}\right)+\frac{p^2}{3}\,F_1^2\left(\frac{\xi_0}{l}\right)K'\right]\frac{T_c-T}{T_c}+\cdots.$$

The justification of this equation is, however, not quite clear. If, in the dirty limit, the magnetic field $H_{c2}$ is explicitly introduced, we obtain a relation,

$$\frac{H_{c3}(T)}{1.695\,H_{c2}(T)} = 1 + p\,\frac{3}{16}\,\frac{C}{\varepsilon^{(0)\frac{1}{2}}}\,[2\,e\,H_{c2}(T)\,l^2]^{\frac{1}{2}} \qquad (21.49)$$

$$+\left[\frac{3}{5}\left(K-p'\,\frac{C}{3\,\varepsilon^{(0)\frac{1}{2}}}\right)+p^2\left(\frac{3}{16}\right)^2 K'\right]2\,e\,H_{c2}(T)\,l^2+\cdots.$$

This equation is correct for all temperatures smaller than $T_c$. It may also be obtained with the help of the method presented at the end of Section 17c. Results similar to Eqs. (21.45), (21.48), and (21.49), though not yet quite correct, were obtained by *Usadel* and *Schmidt* [14].

Calculations of $H_{c3}$ for a clean metal with specularly reflecting surface were done by *Hu* and *Korenman* [32, 59] both for $T \lesssim T_c$ and for $T$ close to zero. The results for temperatures close to zero meet perhaps with some doubt since the authors neglected the curvature of the orbits and the Landau quantization. For $T$ close to $T_c$, they indicate an approximation scheme which is different from the one used here. Whereas the general idea of Section 15 was to extrapolate a solution to the GL equation through a surface layer where it is not valid, in the *Hu* and *Korenman* scheme it is tried to take the pair potential in the surface layer more accurately into account. A similar scheme was subsequently developed by *Sarma* [58]. His result is identical with Eq. (21.45) up to the order $(T_c - T)/T_c$. Deviations, which should occur in higher order, are not yet known.

## c) General Theorems Concerning the Electric Current Density

As shown in Section 21a on the basis of Eq. (21.8), the electric current density has only one non-vanishing component, the $y$ component, in the present geometry as long as the GL or diffusion approximations are valid. Further, the net current vanishes,

$$\int_0^\infty i_y(x)\,\mathrm{d}x = 0\,; \tag{21.50}$$

cf. Eq. (21.10). It shall now be shown that these assertions are true under very general conditions [5].

As pointed out in the beginning of Section 21 b, the ansatz (19.3) may still be used. Defining the kernel

$$K_\omega(x, x'; k, \mu) = \int_{-\infty}^{+\infty} K_\omega(r, r')\exp[ik(y'-y)+i\mu(z'-z)]\,\mathrm{d}y'\mathrm{d}z', \tag{21.51}$$

we may write an integral equation for the function $f(x)$,

$$f(x) = g\,T\sum_\omega{}'\int_0^\infty K_\omega(x, x'; k, \mu)\,f(x')\,\mathrm{d}x'. \tag{21.52}$$

This equation is equivalent to the gap equation (10.1). The real functional

$$\varphi = \frac{1}{g}\int_\sigma^\infty |f(x)|^2\,\mathrm{d}x - \int_0^\infty f^*(x)\,T\sum_\omega{}' K_\omega(x, x'; k, \mu)\,f(x')\,\mathrm{d}x\,\mathrm{d}x' \tag{21.53}$$

vanishes and is stationary with respect to arbitrary variations of $f(x)$, if Eq. (21.52) is satisfied. The functional $\varphi$ follows from the functional $\Phi$, given in Eq. (2.20), if the $y$ and $z$ integrations are not carried out there.

The number of zeros of $f(x)$ for $x \geq 0$ and the parameters $k$ and $\mu$ have to be chosen in such a way that the magnetic field becomes maximal. This field is the critical field $H_{c3}$. It is likely that the number of zeros has again to be chosen equal to zero as in Section 21 a and that also $\mu$ has to vanish. But the optimal value of $k$ has still to be determined. Since $\varphi = 0$ relates $H$ and $k$, we require

$$\int_0^\infty f^*(x)\,T\sum_\omega{}'\frac{\partial K_\omega(x, x'; k, \mu)}{\partial k}\,f(x')\,\mathrm{d}x\,\mathrm{d}x' = 0. \tag{21.54}$$

The contributions from $\partial f(x)/\partial k$ vanish because of the stationarity of $\varphi$. We want to show that Eq. (21.50) is a consequence of Eq. (21.54).

Let us introduce the distribution function

$$\begin{aligned}
&\Delta_\omega(x, v; k, \mu)\\
&\quad = \int g_\omega(r, v; r')\,f(x')\exp[ik(y'-y)+i\mu(z'-z)]\,\mathrm{d}x'\mathrm{d}y'\mathrm{d}z'.
\end{aligned} \tag{21.55}$$

Eq. (21.54) may then be written in the form

$$\int_0^\infty \mathrm{d}x\oint \mathrm{d}\Omega\, f^*(x)\,T\sum_\omega{}'\frac{\partial \Delta_\omega(x, v; k, \mu)}{\partial k} = 0 \tag{21.56}$$

with the understanding that $f(x')$ in Eq. (21.55) stays undifferentiated.

The distribution function $\Delta_\omega(x, v; k, \mu)$ satisfies the Boltzmann equation

$$\left(2|\omega| + v_x \frac{\partial}{\partial x} + iv_y(k + 2eHx) + iv_z\mu + nv\sigma\right)\Delta_\omega(x, v; k, \mu)$$

$$- nv \oint \frac{d\sigma(\theta)}{d\Omega} \Delta_\omega(x, v'; k, \mu) \, d\Omega' = \frac{1}{(2\pi)^2} f(x) \tag{21.57}$$

for $x \geq 0$, and appropriate semi-microscopic boundary conditions at $x = 0$. Differentiation with respect to $k$ [cf. remark after Eq. (21.56)] results in

$$\left(2|\omega| + v_x \frac{\partial}{\partial x} + iv_y(k + 2eHx) + iv_z\mu + nv\sigma\right)\frac{\partial\Delta_\omega(x, v; k, \mu)}{\partial k}$$

$$- nv \oint \frac{d\sigma(\theta)}{d\Omega} \frac{\partial\Delta_\omega(x, v'; k, \mu)}{dk} \, d\Omega' = -iv_y\Delta_\omega(x, v; k, \mu). \tag{21.58}$$

This differential equation plus boundary condition at $x = 0$, may be solved with the help of the Green's function $e_\omega(x, v; x'', v''; k, \mu)$, which satisfies Eq. (21.57) with the righthand side replaced by $\delta(x - x'')\,\hat\delta(v, v'')$. Obviously, it is

$$\Delta_\omega(x, v; k, \mu) = \frac{1}{(2\pi)^2} \int_0^\infty dx' \oint d\Omega' e_\omega(x, v; x', v'; k, \mu) f(x') \tag{21.59}$$

and

$$\frac{\partial\Delta_\omega(x, v; k, \mu)}{\partial k} = \int_0^\infty dx' \oint d\Omega' e_\omega(x, v; x', v'; k, \mu)(-iv_y')\Delta_\omega(x', v'; k, \mu). \tag{21.60}$$

Further it is

$$e_\omega(x, v; x', v'; k, \mu) = e_\omega^*(x', -v'; x, -v; k, \mu), \tag{21.61}$$

which is proved in the same way as Eq. (16.11). Inserting Eq. (21.60) into Eq. (21.56) and making use of Eqs. (21.61) and (21.59), we obtain

$$T \sum_\omega{}' \int_0^\infty dx' \oint d\Omega' \Delta_\omega^*(x', -v'; k, \mu)(-iv_y')\Delta_\omega(x', v'; k, \mu) = 0. \tag{21.62}$$

Since the electric current density may, according to Eq. (18.7), be written as

$$i(x) = -ie(2\pi)^4 N T \sum_\omega{}' \oint d\Omega \, \Delta_\omega^*(x, -v; k, \mu) v \Delta_\omega(x, v; k, \mu), \tag{21.63}$$

Eq. (21.50) immediately follows from Eq. (21.62). Eq. (21.50) was, for specular reflexion, also proved by *Hu* [60] on the basis of a different method.

In the same way it can be shown that optimal choice of $\mu$,

$$\int_0^\infty f^*(x)\, T \sum_\omega{}' \frac{\partial K_\omega(x, x'; k, \mu)}{\partial \mu}\, f(x')\, dx\, dx' = 0 , \qquad (21.64)$$

leads to a vanishing net current in $z$ direction. If, however, the optimal value of $\mu$ is equal to zero and if the reflexion kernel is invariant under the simultaneous substitutions $v_{z\,in} \to - v_{z\,in}$, $v_{z\,out} \to - v_{z\,out}$, it may even be shown that it is

$$i_z(x) = 0 . \qquad (21.65)$$

Under the conditions mentioned, it indeed follows from Eq. (21.57) that it is

$$\Delta_\omega(x, v_x, v_y, v_z; k, \mu = 0) = \Delta_\omega(x, v_x, v_y, -v_z; k, \mu = 0) . \qquad (21.66)$$

Eq. (21.65) is an immediate consequence, if the current density is taken from Eq. (21.63). Also the net current in $z$ direction vanishes, so that Eq. (21.64) is satisfied for $\mu = 0$. But it can, of course, not be shown on the basis of these arguments, that $\mu = 0$ is the optimal choice.

That also $i_x(x)$ vanishes is seen in the following way. According to Eq. (21.63), the current density depends only upon the $x$ coordinate. Eq. (18.12) therefore reduces to

$$\frac{d\,i_x(x)}{dx} = 0 , \qquad (21.67)$$

so that $i_x(x)$ does actually not depend upon $x$. Finally it is

$$i_x(0) = 0 \qquad (21.68)$$

because of Eq. (18.10).

Similar general considerations may also be made in connection with the critical field $H_{c\parallel}$ (Section 22).


## 22. The Surface Parallel Critical Field, $H_{c\,\shortparallel}$, of Layers

We imagine a metallic layer of thickness $a$ with plane parallel surfaces. The surface parallel homogeneous magnetic field, for which a second kind transition between normal and superconducting state occurs, is called $H_{c\parallel}$. For infinite thickness of the layer, we are evidently back to the problem treated in Section 21 so that $H_{c\parallel}$ becomes equal to $H_{c3}$ for $a \to \infty$.

## a) Ginzburg-Landau and Diffusion Approximation

The original calculations were made by *Saint-James* and *de Gennes* [54] in the same paper in which also $H_{c3}$ was calculated. If the two plane surfaces are parallel to the $y - z$ plane of a Cartesian coordinate system and if the magnetic field is parallel to the $z$ direction, the gauge (19.1) may again be used. The GL equation is given by Eq. (19.2). Again the ansatz (19.3) may be made where $f(x)$ satisfies Eq. (19.5). Now a boundary condition of the form of Eq. (21.2) has to be postulated on both surfaces. It turns out that $\mu$ has again to be put equal to zero.

It is again not only required that the temperature is close to the transitions temperature, $T_c$, (GL approximation) or that the electronic mean free path is sufficiently short (diffusion approximation). Also the linear dimensions of the sample (i. e., its thickness, $a$) have to be sufficiently big. The more quantitative formulation, given at the end of Section 20a, is also valid in the present case.

For the explicit calculations we introduce the dimensionless variables of Eq. (21.13). We further put

$$f(x) =: u(\xi) . \tag{22.1}$$

The function $u(\xi)$ satisfies the differential equation

$$\left( \frac{\partial^2}{\partial \xi^2} - \xi^2 \right) u(\xi) = -\varepsilon u(\xi) . \tag{22.2}$$

Since $k$ was eliminated from the formulation, we only have to postulate that it is

$$\frac{\partial u(\xi)}{\partial \xi} = 0 \tag{22.3}$$

at two points, $\xi_r$ and $\xi_l$ (right and left), for which it is

$$\delta := \xi_r - \xi_l = a \sqrt{2eH} . \tag{22.4}$$

Incidentally, it is

$$\varepsilon \delta^2 = \lambda(T) a^2 . \tag{22.5}$$

If the original coordinate system was chosen in such a way that the two surfaces are situated at $x = \pm a/2$, the value of $k$ in the ansatz (19.3) is obtained from

$$\bar{\xi} := \frac{\xi_r + \xi_l}{2} = \frac{k}{\sqrt{2eH}} . \tag{22.6}$$

Following *Schultens* [61], we introduce a basis, $u_+ (\varepsilon, \xi)$ and $u_- (\varepsilon, \xi)$, of solutions to Eq. (22.2), which are defined by

$$u_+ (\varepsilon, 0) = u'_- (\varepsilon, 0) = 1 , \qquad u'_+ (\varepsilon, 0) = u_- (\varepsilon, 0) = 0 . \tag{22.7}$$

The dash denotes the derivative with respect to $\xi$. $u_+(\varepsilon, \xi)$ is an even function and $u_-(\varepsilon, \xi)$ an odd function, of $\xi$. The Wronskian,

$$u_+(\varepsilon, \xi)\, u'_-(\varepsilon, \xi) - u'_+(\varepsilon, \xi)\, u_-(\varepsilon, \xi) = 1 \,, \tag{22.8}$$

is independent of $\xi$. The general solution to Eq. (22.2) may now be written as linear combination,

$$u(b, \varepsilon, \xi) = u_+(\varepsilon, \xi) + b u_-(\varepsilon, \xi)\,, \tag{22.9}$$

of the basis. When the odd function,

$$\varphi(\varepsilon, \xi) := \frac{u'_+(\varepsilon, \xi)}{u'_-(\varepsilon, \xi)} \tag{22.10}$$

is introduced, the boundary condition (22.3) can be satisfied for every pair of $\xi$ values, $\xi_l$ and $\xi_r$, for which it is

$$\varphi(\varepsilon, \xi_l) = \varphi(\varepsilon, \xi_r)\,. \tag{22.11}$$

The constant, $b$, in Eq. (22.9) has simply to be put equal to the negative of this common value. The function $\varphi(\varepsilon, \xi)$ is shown graphically in Fig. 1 of Ref. [61].

Schultens [61] uses a different notation: his $u$ is our $\xi$, his $E$ is our $\varepsilon$.

Eq. (22.11) is an implicit relation between magnetic field strength, $H$, and the parameter $k$ [Eq. (19.3)] or $\bar{\xi}$ [Eq. (22.6)]. In order to maximize $H$ and so to make it equal to $H_{c\parallel}$, we have to postulate that Eq. (22.11) is stationary with respect to variations of $\bar{\xi}$,

$$\varphi'(\varepsilon, \xi_l) = \varphi'(\varepsilon, \xi_r)\,. \tag{22.12}$$

Eq. (22.12) may also be written in the form

$$(\xi_l^2 - \varepsilon)\, u^2(b, \varepsilon, \xi_l) = (\xi_r^2 - \varepsilon) u^2(b, \varepsilon, \xi_r)\,, \tag{22.13}$$

as follows from Eqs. (22.10), (22.2), (22.8), and from

$$u(b, \varepsilon, \xi_l)\, u'_-(\varepsilon, \xi_l) = 1 = u(b, \varepsilon, \xi_r)\, u'_-(\varepsilon, \xi_r)\,, \tag{22.14}$$

which itself is a consequence of Eqs. (22.9) and (22.8) and of the choice of $b$ explained in connection with Eq. (22.11).

When the program is carried through, it is seen that all relevant values of $\varepsilon$ lie in the interval

$$0 < \varepsilon < \varepsilon^{(0)} \tag{22.15}$$

with $\varepsilon^{(0)}$ given by Eq. (21.19). For

$$0 < \varepsilon \leq 0.505\,, \tag{22.16}$$

it is $\bar{\xi} = 0$ and

$$\xi_r = -\xi_l = \frac{\delta}{2}. \tag{22.17}$$

It is $b$ equal to zero so that the function $u(b, \varepsilon, \xi)$ becomes equal to the even function $u_+(\varepsilon, \xi)$. In the remaining interval,

$$0.505 < \varepsilon < \varepsilon^{(0)}, \tag{22.18}$$

$b$ and $\bar{\xi}$ are different from zero. There are now two linearly independent solutions with opposite values of $b$ and $\bar{\xi}$, and so of $k$.

The limiting cases are simple. For $\varepsilon \to \varepsilon^{(0)}$ or $\delta \to +\infty$, $\xi_l$ approaches $-\varepsilon^{(0)\frac{1}{2}}$ and $\xi_r$ approaches $+\infty$ (or $\xi_l \to -\infty$ and $\xi_r \to +\varepsilon^{(0)\frac{1}{2}}$). Obviously, $H_{c\parallel}$ approaches the critical field $H_{c3}$ (Section 21),

$$\frac{H_{c\parallel}(T)}{H_{c2}(T)} \to 1.695. \tag{22.19}$$

In terms of the original units, this limit may be characterized as

$$a \gg \xi(T) \tag{22.20}$$

with $\xi(T)$ defined in Eq. (13.2).

The opposite limit, $\varepsilon \to 0$, $\delta \to 0$, still requires some more detailed calculations. It follows from Eqs. (22.2) and (22.7), that it is

$$u_+(\varepsilon, \xi) = 1 - \frac{\varepsilon}{2}\xi^2 + \frac{1}{12}\xi^4 + \cdots, \quad u_-(\varepsilon, \xi) = \xi - \frac{\varepsilon}{3}\xi^3 + \cdots. \tag{22.21}$$

Since $u(b, \varepsilon, \xi)$ now equals $u_+(\varepsilon, \xi)$, Eq. (22.3) leads to

$$\xi_r = -\xi_l = \sqrt{3\varepsilon} \tag{22.22}$$

or

$$\delta = \sqrt{12\varepsilon}. \tag{22.23}$$

Combination with Eqs. (22.5), (21.13), (13.2), and (19.8) gives the result

$$2eH_{c\parallel}(T) = \frac{\sqrt{12}}{\xi(T)a} = \frac{\sqrt{24eH_{c2}(T)}}{a}. \tag{22.24}$$

If $\hbar$ and $c$ are not put equal to one, this result may also be written as

$$H_{c\parallel}(T) = \frac{1}{a}\sqrt{\frac{6}{\pi}H_{c2}(T)\,\Phi_0}, \tag{22.25}$$

where $\Phi_0$ is the unit of quantized flux given after Eq. (19.9). The present limit, $\varepsilon \ll 1$, may roughly be written as

$$a \ll \xi(T). \tag{22.26}$$

We also list, for future application,

$$\varphi(\varepsilon, \xi) = -\varepsilon\xi + \frac{1}{3}\,\xi^3 + \cdots,\qquad (22.27)$$

which follows from Eqs. (22.10) and (22.21).

Returning now to the general problem, we see that the electric current density is still given by Eq. (21.8). The net current again vanishes,

$$\int_{\xi_l}^{\xi_r} \xi u^2(\xi)\,\mathrm{d}\xi = 0. \qquad (22.28)$$

This follows from similar considerations as those made at the end of Section 21a. Eqs. (22.3) and (22.12) are decisive for the proof.

In the interval (22.18), it seems unphysical to work with the simple solution derived above. Instead, a linear combination of the two solutions with opposite values of $b$ (and of $\bar{\xi}$) has to be formed. Though it cannot be decided from the linear GL equation which combination is the right one, it seems plausible that both solutions have to be combined with equal weight factors. The current pattern becomes then more complicated. Current eddies appear. The details were given in Ref. [61].

### b) Going beyond the Approximations

For a clean metal, boundary condition (22.3) is now replaced by

$$u'(\xi_l) = \begin{cases} -p\,\dfrac{45\zeta(4)}{224\zeta(3)}\,\eta\,\xi_l^2 u(\xi_l), \\[2ex] -\dfrac{31\zeta(5)}{70\zeta(3)}\,\eta^2\,\xi_l u(\xi_l), \end{cases} \qquad u'(\xi_r) = \begin{cases} p\,\dfrac{45\zeta(4)}{224\zeta(3)}\,\eta\,\xi_r^2 u(\xi_r), \\[2ex] -\dfrac{31\zeta(5)}{70\zeta(3)}\,\eta^2\,\xi_r u(\xi_r). \end{cases} \qquad (22.29)$$

The respective first line is valid for non-specular reflexion, and it is assumed that the diffuseness, $p$, is the same on both surfaces. The second line is valid for specular reflexion. Eq. (22.29) is closely related to Eq. (21.21). Both equations are consequences of Eq. (15.32). Starting from the solution derived in Section 22a we have to alter $b, \varepsilon, \xi_l$, and $\xi_r$ in such a way that Eq. (22.29) is satisfied and that Eq. (22.5) is still fulfilled. This leads to an alteration of $H_{c\parallel}(T)$ which shall now be calculated [62].

It is convenient to introduce the function

$$W(\varepsilon, \bar{\xi}, \delta) := \varphi(\varepsilon, \xi_l) - \varphi(\varepsilon, \xi_r) \qquad (22.30)$$

with $\varphi(\varepsilon, \xi)$ given in Eq. (22.10). Eqs. (22.11) and (22.12) may be written in the form

$$W(\varepsilon, \bar{\xi}, \delta) = 0 = \frac{\partial W(\varepsilon, \bar{\xi}, \delta)}{\partial \bar{\xi}}. \qquad (22.31)$$

For any variation of the parameters about values satisfying Eq. (22.31), it is

$$d\,W(\varepsilon, \overline{\xi}, \delta) = u(\xi_l)\,d\,u'(\xi_l) - u(\xi_r)\,d\,u'(\xi_r)\,,\tag{22.32}$$

where $u(\xi) = u(b, \varepsilon, \xi)$ is the solution (22.9) with

$$b = -\varphi(\varepsilon, \xi_l) = -\varphi(\varepsilon, \xi_r)\,.\tag{22.33}$$

Eq. (20.32) is a consequence of Eqs. (22.9) and (22.14). Contributions from the variation of $b$ to the righthand side vanish because of Eq. (22.14). On the lefthand side, $\varepsilon$ and $\delta$ may be varied,

$$d\,W = \frac{\partial W}{\partial \varepsilon}\,d\varepsilon + \frac{\partial W}{\partial \delta}\,d\delta\,;\tag{22.34}$$

cf. Eq. (22.31).

Introducing the variable $\eta$ from Eq. (21.22) with

$$d\delta = \frac{\delta}{\eta}\,d\eta\,,\tag{22.35}$$

we obtain the equation

$$\frac{\partial W}{\partial \varepsilon}\,d\varepsilon + \frac{\partial W}{\partial \delta}\,\frac{\delta}{\eta}\,d\delta = B\,,\tag{22.36}$$

with

$$B = \begin{cases} -p\,\dfrac{45\zeta(4)}{224\zeta(3)}\,\eta\,[\xi_l^2\,u^2(\xi_l) + \xi_r^2\,u^2(\xi_r)]\,, \\[2mm] -\dfrac{31\zeta(5)}{70\zeta(3)}\,\eta^2\,[\xi_l\,u^2(\xi_l) - \xi_r\,u^2(\xi_r)]\,. \end{cases}\tag{22.37}$$

The first line holds for non-specular reflexion, the second one for specular reflexion. Eq. (22.37) is a consequence of Eqs. (22.32) and (22.29).

Eq. (22.36) is one linear equation for $d\varepsilon$ and $d\delta$. The other one is obtained from Eq. (21.23),

$$\eta^2\,d\varepsilon + 2\eta\varepsilon\,d\eta = A\,.\tag{22.38}$$

The quantity $A$ may be put equal to zero for $p \neq 0$. It suffices to use

$$A = \frac{93\zeta(5)}{140\zeta(3)}\,\eta^4 + \cdots\tag{22.39}$$

for $p = 0$ and $a/2 \ll \xi(T)$. Since we are interested in the magnetic field, $H_{c\parallel}(T)$, Eqs. (22.36) and (22.38) are now solved for $d\eta$ with the result

$$d\eta = \frac{A\,\dfrac{\partial W}{\partial \varepsilon} - B\eta^2}{\eta\left(2\varepsilon\,\dfrac{\partial W}{\partial \varepsilon} - \delta\,\dfrac{\partial W}{\partial \delta}\right)}\,.\tag{22.40}$$

Evidently it is

$$H_{c\parallel}(T) = H_{c\parallel}^{(0)}(T)\left(1 + \frac{2d\eta}{\eta} + \cdots\right), \tag{22.41}$$

where $H_{c\parallel}^{(0)}(T)$ is the result of Sec. 22a.

In the case of Eq. (22.20), the results of Section 21b, in particular Eqs. (21.28) and (21.29), are again obtained. Therefore, we shall only study the opposite limit (22.26). It follows from Eqs. (22.30), (22.27), (22.4), and (22.6), that it is

$$W = \varepsilon\delta - \bar{\xi}^2\delta - \frac{\delta^3}{12} + \cdots \tag{22.42}$$

and, therefore,

$$\frac{\partial W}{\partial \varepsilon} = \delta, \quad \frac{\partial W}{\partial \delta} = \varepsilon - \frac{\delta^2}{4} = -2\varepsilon. \tag{22.43}$$

In deriving the last equation, $\bar{\xi}$ was put equal to zero and Eq. (22.23) was used. Further, it is

$$B = \begin{cases} -p\,\dfrac{45\zeta(4)}{224\zeta(3)}\,\eta\,\dfrac{\delta^2}{2}, \\[2mm] \dfrac{31\zeta(5)}{70\zeta(3)}\,\eta^2\delta \end{cases} \tag{22.44}$$

as a consequence of Eqs. (22.37), (22.17), and (22.21). In this way, we obtain

$$H_{c\parallel}(T) = \frac{1}{a}\sqrt{\frac{6}{\pi}\,H_{c2}(T)\,\Phi_0}\left(1 + \begin{cases} p\,\dfrac{135\zeta(4)}{224\zeta(3)}\,\dfrac{\xi_0(T_c)}{a} + \cdots, \\[2mm] \dfrac{93\zeta(5)}{70\zeta(3)}\left(\dfrac{\xi_0(T_c)}{a}\right)^2 + \cdots \end{cases}\right) \tag{22.45}$$

from Eqs. (22.41), (22.40), and (22.25).

The correction terms in Eq. (22.45) are temperature independent like those in Eq. (20.11). They arise from the fact, that all linear dimensions of a sample (including its thickness) have to be big as compared to $\xi_0(T_c)$, the thickness of the surface layer discussed in Section 15a. The correction term is one order of magnitude smaller for $p = 0$. The reason for this difference was already pointed out in connection with Eqs. (21.28) and (21.29).

The calculations may be extended to samples containing impurities provided that their concentration is homogeneous and that the differential cross section is isotropic. The calculations start from Eqs. (17.39) and

(17.47) respectively. In these equations, it is

$$\frac{d\lambda}{dH} = 2e\,\frac{2\varepsilon\,\dfrac{\partial W}{\partial \varepsilon} - \delta\,\dfrac{\partial W}{\partial \delta}}{2\,\dfrac{\partial W}{\partial \varepsilon}}, \qquad (22.46)$$

since Eq. (22.31) defines $\lambda$ as a function of $H$. Eqs. (22.34), (22.4), and (21.13) were used in deriving Eq. (23.46). In Eq. (17.39) there further occurs

$$\frac{\oint p|n \times \tilde{\partial}\varDelta(r)|^2\,|d^2r|}{\int |\varDelta(r)|^2\,d^3 r} = \frac{(2eH)^{\frac{1}{2}}p[\xi_l^2 u^2(\xi_l) + \xi_r^2 u^2(\xi_r)]}{\displaystyle\int_{\xi_l}^{\xi_r} u^2(\xi)\,d\xi}, \qquad (22.47)$$

where it is

$$\int_{\xi_l}^{\xi_r} u^2(\xi)\,d\xi = u(\xi_l)\,\frac{\partial}{\partial \varepsilon}\,u'(\xi_l) - u(\xi_r)\,\frac{\partial}{\partial \varepsilon}\,u'(\xi_r) = \frac{\partial W}{\partial \varepsilon}. \qquad (22.48)$$

The first part of the latter equation is derived in a similar way as Eq. (21.35). Eq. (22.32) was used for the second equality sign.

The final result is

$$H_{c\parallel}(T) = \frac{1}{a}\sqrt{\frac{6}{\pi}}\,H_{c2}(T)\,\varPhi_0\left(1 + \left\{\begin{array}{l} 3p\,F_1\!\left(\dfrac{\xi_0}{l}\right)\dfrac{\xi_0(T_c)}{a} + \cdots \\[2mm] 6\,F_2\!\left(\dfrac{\xi_0}{l}\right)\!\left(\dfrac{\xi_0(T_c)}{a}\right)^{\!2} + \cdots \end{array}\right.\right), \qquad (22.49)$$

where the first line holds for non-specular, the second one for specular reflexion. The clean limit result (22.45) is again obtained with the help of Eqs. (17.42) and (17.50). The dirty limit result,

$$H_{c\parallel}(T) = \frac{1}{a}\sqrt{\frac{6}{\pi}}\,H_{c2}(T)\,\varPhi_0\left(1 + \left\{\begin{array}{l} \dfrac{9}{16}\,p\,\dfrac{l}{a} + \cdots \\[2mm] \dfrac{6}{5}\left(\dfrac{l}{a}\right)^{\!2} + \cdots \end{array}\right.\right), \qquad (22.50)$$

is obtained if Eqs. (17.43) and (17.51) are applied. Whereas Eq. (22.49) is in general only valid for temperatures, $T$, close to $T_c$, Eq. (22.50) should be valid at all temperatures smaller than $T_c$.

Also the correction terms in Eqs. (22.49) and (22.50) are temperature independent. They originate from the requirement that the film thickness has to be sufficiently thick in order that GL approximation or diffusion approximation are valid.

## c) Thin Films

A special method shall now be developed, which is applicable to layers and film thin enough so that the pair potentials is approximately constant in space. Such a situation has already been encountered in connection with Eq. (22.24) and with the various corrections to this equation derived in Section 22b. Those results will actually be rederived. But the present method may also be applied to films which are too thin for the GL approximation and diffusion approximation to be valid.

The statement, that the pair potential is constant in space, is evidently not gauge invariant. We shall assume that the gauge to be used is the one given by Eq. (19.1) if the film is extended from $x = -a/2$ to $x = +a/2$. This assumption is suggested by the symmetry of the problem. It leads to the correct result in the GL and diffusion approximations.

We use $K_\omega(x, x')$ as a shorthand notation for $K_\omega(x, x'; 0, 0)$ defined by Eq. (21.51) and introduce a quantity $\kappa_\omega$ by

$$\kappa_\omega = \frac{v}{2\pi N a} \int_{-a/2}^{+a/2} K_\omega(x, x') \, dx \, dx'. \tag{22.51}$$

It shall be shown below in small print, that Eqs. (17.25) and (19.20) are valid in this case though $\kappa_\omega$ is now defined, not as an eigenvalue, but as an expectation value. For isotropically scattering impurities with homogeneous concentration, we may further define

$$\kappa_\omega^{(\sigma)} = \frac{v}{2\pi N a} \int_{-a/2}^{+a/2} K_\omega^{(\sigma)}(x, x') \, dx \, dx'. \tag{22.52}$$

It shall also be shown below that Eq. (17.9) is still valid.

In order to derive Eq. (17.25), we start from the variational principle based upon the functional $\varphi$ as given in Eq. (21.53). The integrations are now, however, extended only over the interval $-a/2 \le x \le +a/2$. $f(x)$ is put constant in Eq. (21.53). Varying the value of the constant leads to $\varphi = 0$, which in turn gives Eq. (17.25). Further, in order to justify Eq. (17.9), we start [63] from Eq. (17.3) or, with adaption to the present problem, from

$$K_\omega(x, x') = K_\omega^{(\sigma)}(x, x') + \frac{v}{2\pi N l} \int_{-a/2}^{+a/2} K_\omega^{(\sigma)}(x, x'') K_\omega(x'', x) \, dx''. \tag{22.53}$$

This equation is the stationarity condition of the functional

$$J_\omega = \int_{-a/2}^{+a/2} \Delta^*(x) [K_\omega(x, x') K_\omega(x', x'') - K_\omega(x, x') K_\omega^{(\sigma)}(x', x'') - K_\omega^{(\sigma)}(x, x') K_\omega(x', x'')] \Delta(x'') \, dx \, dx' \, dx''$$

$$- \frac{v}{2\pi N l} \int_{-a/2}^{+a/2} \Delta^*(x) K_\omega(x, x') K_\omega^{(\sigma)}(x', x'') K_\omega(x'', x''') \Delta(x''') \, dx \, dx' \, dx'' \, dx''', \tag{22.54}$$

where $\Delta(x)$ is an arbitrary but given function. $J_\omega$ is closely related to the functional given in Eq. (16.13). Now we put $\Delta(x)$ equal to one and take

$$\int_{-a/2}^{+a/2} K_\omega(x', x'') \Delta(x'') \, dx'' = \frac{2\pi N}{v} \kappa_\omega \tag{22.55}$$

as an $x'$ independent test function. With this ansatz, Eq. (22.54) reads

$$J_\omega = a \left( \frac{2\pi N}{v} \right)^2 \left[ |\kappa_\omega|^2 - \kappa_\omega^* \kappa_\omega^{(\sigma)} - \kappa_\omega^{(\sigma)*} \kappa_\omega - \frac{1}{l} \kappa_\omega^* \kappa_\omega^{(\sigma)} \kappa_\omega \right], \qquad (22.56)$$

where $\kappa_\omega^{(\sigma)}$ is defined by Eq. (22.52). Variation with respect to $\kappa_\omega^*$ immediately leads to Eq. (17.9).

The kernel $K_\omega^{(\sigma)}(x, x')$ appearing above in Eq. (22.52) may be derived from a distribution function,

$$K_\omega^{(\sigma)}(x, x') = 2\pi^2 N \oint g_\omega^{(\sigma)}(x, v; x') \, d\Omega, \qquad (22.57)$$

which satisfies the Boltzmann equation

$$\left( 2|\omega| + v_x \frac{\partial}{\partial x} + 2ieHxv_y + nv\sigma \right) g_\omega^{(\sigma)}(x, v; x') = \frac{1}{(2\pi)^2} \delta(x - x') \qquad (22.58)$$

[cf. Eq. (12.11)] and semi-microscopic boundary conditions at $x = \pm a/2$. For vanishing magnetic field ($H = 0$), it follows

$$\int_{-a/2}^{+a/2} g_\omega^{(\sigma)}(x, v; x') \, dx' = \frac{1}{(2\pi)^2 (2|\omega| + nv\sigma)} = \frac{\zeta_\omega}{(2\pi)^2 v} \qquad (22.59)$$

from Eq. (22.58) and from property (10.20) of the reflexion kernel. Also Eq. (12.14) was used. Eq. (22.52) then leads to

$$\kappa_\omega^{(\sigma)}(H = 0) = \zeta_\omega. \qquad (22.60)$$

If a Taylor expansion in terms of $H$ is made about $H = 0$,

$$\kappa_\omega^{(\sigma)}(H) = \kappa_\omega^{(\sigma)}(0) + H \frac{d\kappa_\omega^{(\sigma)}(0)}{dH} + \frac{H^2}{2} \frac{d^2 \kappa_\omega^{(\sigma)}(0)}{dH^2} + \cdots, \qquad (22.61)$$

the first term is given by Eq. (22.60). The second term vanishes as shall be shown below. Therefore, we are left with

$$\kappa_\omega^{(\sigma)}(H) = \zeta_\omega + \frac{H^2}{2} \frac{d^2 \kappa_\omega^{(\sigma)}}{dH^2} + \cdots. \qquad (22.62)$$

Terms of higher than the second order in the magnetic field shall be neglected.

Our next task is to calculate the quantity

$$\frac{d^2 \kappa_\omega^{(\sigma)}(0)}{dH^2} = \frac{v\pi}{a} \int_{-a/2}^{+a/2} dx \, dx' \oint d\Omega \frac{\partial^2 g_\omega^{(\sigma)}(x, v; x')}{\partial H^2}. \qquad (22.63)$$

The righthand side is a consequence of Eqs. (22.52) and (22.57). It now follows from Eq. (22.58) by differentiation with respect to $H$, that it is

$$\left( \frac{v}{\zeta_\omega} + v_x \frac{\partial}{\partial x} + 2ieHxv_y \right) \frac{\partial g_\omega^{(\sigma)}(x, v; x')}{\partial H} = -2iexv_y g_\omega^{(\sigma)}(x, v; x') \qquad (22.64)$$

and

$$\left(\frac{v}{\zeta_\omega} + v_x \frac{\partial}{\partial x} + 2ieHxv_y\right) \frac{\partial^2 g_\omega^{(\sigma)}(x, v; x')}{\partial H^2} = -4iexv_y \frac{\partial g_\omega^{(\sigma)}(x, v; x')}{\partial H}.$$

$$(22.65)$$

The length $\zeta_\omega$ was again introduced from Eq. (12.14). The semi-microscopic boundary conditions are also valid for the derivatives. Since these derivatives have to be taken at $H = 0$, the magnetic field, $H$, has to be put equal to zero on the lefthand sides of Eqs. (22.64) and (22.65). Integrating Eq. (22.64) with respect to $x'$ and making use of Eq. (22.59), we obtain the result that it is

$$\int_{-a/2}^{+a/2} \frac{\partial g_\omega^{(\sigma)}(x, v; x')}{\partial H} dx' = -\frac{2ie\zeta_\omega}{(2\pi)^2} f_\omega(x, v),$$

$$(22.66)$$

where the distribution function, $f_\omega(x, v)$, satisfies the Boltzmann equation

$$\left(\frac{v}{\zeta_\omega} + v_x \frac{\partial}{\partial x}\right) f_\omega(x, v) = x\frac{v_y}{v},$$

$$(22.67)$$

and semi-microscopic boundary conditions at $x = \pm a/2$. It further follows from Eq. (22.65) that it is

$$\int_{-a/2}^{+a/2} \frac{\partial^2 g_\omega^{(\sigma)}(x, v; x')}{\partial H^2} dx = -4ie\zeta_\omega \frac{v_y}{v} \int_{-a/2}^{+a/2} \frac{\partial g_\omega^{(\sigma)}(x, v; x')}{\partial H} x\, dx.$$

$$(22.68)$$

The boundary terms vanish because of Eq. (10.21). Inserting Eq. (22.66) into Eq. (22.68) and the latter into Eq. (22.63), we obtain

$$\frac{d^2 \kappa_\omega^{(\sigma)}(0)}{dH^2} = -(2e)^2 \frac{\zeta_\omega^2}{2\pi a} \int_{-a/2}^{+a/2} dx \oint d\Omega\, x v_y f_\omega(x, v).$$

$$(22.69)$$

This expression has now to be further evaluated.

In order to do this, we study the solution to Eq. (22.67). For $v_x > 0$ (motion from "left" to "right"), $f_\omega(x, v)$ is determined by its value, $f_\omega(-a/2, v)$, on the "left" surface ($x = -a/2$) and by a source term,

$$f_\omega(x, v_x, v_\parallel) \qquad\qquad\qquad (22.70)$$

$$= f_\omega(-a/2, v_x, v_\parallel) \exp\left[-\frac{(x+a/2)v}{\zeta_\omega v_x}\right] + \frac{v_y}{v_x v} \int_{-a/2}^{x} \exp\left[-\frac{(x-x')v}{\zeta_\omega v_x}\right] x'\, dx'.$$

Here, the velocity components $v_y$ and $v_z$ have been combined to $v_\parallel$. In the same way, for a motion from right to left, we find

$$f_\omega(x, -v_x, v_\parallel) \qquad\qquad\qquad (22.71)$$

$$= f_\omega(+a/2, -v_x, v_\parallel) \exp\left[\frac{(x-a/2)v}{\zeta_\omega v_x}\right] + \frac{v_y}{v_x v} \int_{x}^{a/2} \exp\left[\frac{(x-x')v}{\zeta_\omega v_x}\right] x'\, dx'$$

with $v_x > 0$. From now on, the reflexion kernel shall be given in the form of Eq. (10.24) with the diffuseness, $p$, being the same on both surfaces. Then it is

$$f_\omega(-a/2, \quad v_x, v_{||}) = (1-p) f_\omega(-a/2, -v_x, v_{||}),$$
$$f_\omega(+a/2, -v_x, v_{||}) = (1-p) f_\omega(+a/2, \quad v_x, v_{||}), \tag{22.72}$$

where it has already been anticipated that the diffuse part of the reflexion kernel does not contribute to $f_\omega(-a/2, v_x, v_{||})$ and $f_\omega(+a/2, -v_x, v_{||})$.

Putting $x$ equal to $+a/2$ in Eq. (22.70) and equal to $-a/2$ in Eq. (22.71) and making use of Eq. (22.72), we obtain a system of linear equations for $f_\omega(a/2, v_x, v_{||})$ and $f_\omega(-a/2, -v_x, v_{||})$. The solution is given by

$$f_\omega(a/2, v_x, v_{||})$$

$$= \frac{1}{1+(1-p) \exp\left[-\dfrac{av}{\zeta_\omega v_x}\right]} \frac{v_y}{v_x v} \int_{-a/2}^{+a/2} \exp\left[-\frac{(a/2-x')v}{\zeta_\omega v_x}\right] x' dx' \tag{22.73}$$

$$= -f_\omega(-a/2, -v_x, v_{||}).$$

Inserting these boundary values into Eqs. (22.70) and (22.71), we obtain the complete distribution function $f_\omega(x, v)$. It is seen that $f_\omega(x, v)$ is a product of $v_y$ and of a function of $v_x$ and $x$. It follows immediately, that the diffuse part of the reflexion kernel does, indeed, not contribute to Eq. (22.72). It also follows with the help of Eq. (22.66) that $d\kappa_\omega^{(\sigma)}(0)/dH$ vanishes. The latter statement is incidentally true for any reflexion kernel which is invariant with respect to rotations about the $x$ axis. If the complete distribution function is inserted into the righthand side of Eq. (22.69), a straightforward calculation leads to the result [63]

$$\frac{d^2\kappa_\omega^{(\sigma)}(0)}{dH^2} = -(2e)^2 \frac{\zeta_\omega^6}{a} \int_0^1 d\beta(1-\beta^2)\beta^3 \left\{ \frac{a^3}{12\zeta_\omega^3\beta^3} - \frac{a^2}{4\zeta_\omega^2\beta^2} + 1 \right.$$

$$\left. - \left(\frac{a}{2\zeta_\omega\beta} + 1\right)^2 \exp\left[-\frac{a}{\zeta_\omega\beta}\right] \right. \tag{22.74}$$

$$\left. + \frac{1-p}{1+(1-p)\exp\left[-\dfrac{a}{\zeta_\omega\beta}\right]} \left[\frac{a}{2\zeta_\omega\beta} - 1 + \left(\frac{a}{2\zeta_\omega\beta} + 1\right)\exp\left[-\frac{a}{\zeta_\omega\beta}\right]\right]^2 \right\},$$

where the auxiliary variable

$$\beta = \frac{v_x}{v} \tag{22.75}$$

was introduced.

First, we study the case

$$a \gg \zeta_0(T). \tag{22.76}$$

Here, $\zeta_0(T)$ [Eq. (17.20)] is the biggest value of $\zeta_\omega$ appearing in Eq. (22.74). The limit (22.76) corresponds to the GL and diffusion approximations treated in Sections 22a and 22b. All terms in Eq. (22.74), which contain exponentials, may now be discarded. If only the highest power in $a/\zeta_\omega$ is kept, we obtain

$$\frac{d^2\kappa_\omega^{(\sigma)}(0)}{dH^2} = -(2e)^2 \frac{\zeta_\omega^3 a^2}{18} + \cdots. \tag{22.77}$$

Inserting the righthand side into Eq. (22.62) and comparing the result with Eq. (17.17), we obtain

$$(2eH_{c\|})^2 \frac{a^2}{12} = \lambda(T) = 2eH_{c2}, \tag{22.78}$$

which is identical with Eq. (22.24). Keeping also the second highest power in $a/\zeta_\omega$, we derive

$$\frac{d^2\kappa_\omega^{(\sigma)}(0)}{dH^2} = -(2e)^2 \frac{\zeta_\omega^3 a^2}{18} \left(1 - \begin{cases} \dfrac{9}{8} p\, \dfrac{\zeta_\omega}{a}\,, \\[2mm] \dfrac{12}{5}\left(\dfrac{\zeta_\omega}{a}\right)^2\,, \end{cases}\right. \tag{22.79}$$

where the first line is valid for non-specular, the second one for specular reflexion. Eq. (22.49) may be obtained from this result.

More interesting is the opposite limit,

$$a \ll \zeta_0(T), \tag{22.80}$$

in which GL and diffusion approximation are certainly not valid. The evaluation of the righthand side of Eq. (22.74) is more complicated in this case. A complete answer has been obtained [63] for $p=0$ (specular reflexion) and $p=1$ (completely diffuse reflexion),

$$\frac{d^2\kappa_\omega^{(\sigma)}(0)}{dH^2} = \begin{cases} -(2e)^2 \dfrac{4}{\pi^4} \zeta_\omega^2 a^3 & (p=0)\,, \\[3mm] -(2e)^2 \dfrac{1}{32} \zeta_\omega^2 a^3 & (p=1)\,. \end{cases} \tag{22.81}$$

This equation has still to be inserted into Eq. (22.62), and the latter into Eqs. (17.9) and (19.20). The result agrees with one obtained earlier by *Shapoval* [17, 18]. The specular reflexion result was also derived by *Thompson* and *Baratoff* [64] and by *Harms* [65]. A compact formulation may also be obtained for an arbitrary value of $p$,

$$\frac{d^2 K_\omega^{(\sigma)'}(0)}{dH^2}$$

(22.81 a)

$$= -(2e)^2 \zeta_\omega^2 a^3 \left[ \frac{1}{32} + \int\limits_0^\infty \frac{d\alpha}{\alpha} \frac{1-p}{1-p+\exp\alpha} \left( \int\limits_{-\frac{1}{2}}^{+\frac{1}{2}} dx \, x \exp[-\alpha x] \right)^2 \right].$$

The function in the square bracket decreases monotonically with increasing value of $p$.

It is seen that $H_{c\parallel}(T) \, a^{\frac{1}{2}}$ is a function of temperature and electronic mean free path. The function depends also upon the value of $p$. The whole approximation is valid only for sufficiently small values of $H$, i. e., in a certain temperature interval below the transition temperature, $T_c$. Conditions for the validity of the approximation were given in Ref. [63].

A complete evaluation may be made in the dirty limit [Eq. (14.15)]. Eq. (22.80) becomes

$$a \ll l \ll \xi_0(T_c),$$

(22.82)

when it is combined with Eq. (14.15). It is presumably difficult to satisfy this condition in an experiment. The quantity $\zeta_\omega$ may be replaced by $l$, both in Eq. (22.81) and in the second term on the righthand side of Eq. (17.17). A comparison between Eqs. (22.62) and (17.17) then leads to

$$H_{c\parallel}(T) = \begin{cases} \sqrt{\dfrac{\pi^3}{12}} \sqrt{\dfrac{H_{c2}(T) \, \Phi_0 l}{a^3}}, & (p = 0), \\[3ex] \sqrt{\dfrac{32}{3\pi}} \sqrt{\dfrac{H_{c2}(T) \, \Phi_0 l}{a^3}}, & (p = 1). \end{cases}$$

(22.83)

Also Eq. (19.8) was used and the unit of quantized flux, $\Phi_0$, was introduced [cf. Eq. (19.9)]. Eq. (22.83) is expected to be valid at all temperatures, $0 \leqq T \leqq T_c$.

Similar considerations were made earlier by *de Gennes* and *Tinkham* [15]. They started from the concept of classical orbits which is actually due to *de Gennes* (cf. Introduction). The results agree in the dirty limit [Eq. (22.25) for $l \ll a$, Eq. (22.83) for $a \ll l$ and $p = 1$], but they disagree in other cases. The de Gennes-Tinkham approximation scheme seems essentially a scheme for small temperatures and not, as the one presented here, for small magnetic fields. A critical discussion of the de Gennes-Tinkham paper was given by *Shapoval* [17].

We wish to thank *H. J. Sommers* for the communication of Eq. (22.81 a) and for discussions on Ref. [15].

## 23. Several Metals in Electric Contact

### a) General Remarks

Only samples consisting of a single metal have so far been studied in this chapter. In the present section, we turn to the problem of samples consisting of several metals in electric contact. An actual experiment is usually concerned with superposed films. The transition temperature in the absence of a magnetic field, and critical fields may be studied for such a sample both experimentally and theoretically.

The analysis shall be restricted to weak-coupling superconductors in which the phonon-mediated electron-electron coupling is characterized by a coupling constant, $g$, and a cut-off frequency, $\omega_D$. It seems to be an open question how so-called normal metals fit into this picture: whether they are actually superconductors with a very small transition temperature ($gN$ still positive but very small) or whether the phonon-mediated interaction is repulsive ($g$ negative).

Within the scope of the method of the correlation function, coupling constant, cut-off frequency, and (for spherical Fermi surface) Fermi velocity, which are usually different in different metals, are assumed constant inside each particular metal. Special considerations are necessary if the cut-off or Debye frequencies of the metals in contact are different. A suggestive symmetric cut-off was introduced in Eq. (10.2).

The kernel, $K_\omega(r, r')$, may still be derived from a distribution function obeying a Boltzmann equation. But this equation has now to be supplemented by boundary conditions, not only on external surfaces, but also on interfaces between different metals. The latter were given in Eq. (10.26). Reflexion and transmission kernels appearing in these boundary conditions, have to satisfy a number of general conditions summarized in Eqs. (10.27) through (10.31).

From this basic formulation, approximate formulations can be derived. The GL equation (13.1) is a differential equation for the pair potential. It is valid for temperatures close to the transition temperature. This approximation is not very interesting for the present problem, since the transition temperatures of all metals in a sample would have to be almost equal. Boundary conditions on interfaces were given, for clean metals, in Eqs. (15.62) and (15.67). The latter, called simplified boundary conditions, should lead to reliable results for sufficiently good electric contact through the interface.

More interesting from a physical point of view is the diffusion approximation which is valid in the dirty limit [Eq. (14.15)]. The diffusion equation (14.10) is a differential equation for the kernel, $K_\omega(r, r')$. It cannot be transformed into a differential equation for the pair potential, if the sample does not consist of a single metal. Boundary conditions on interfaces were given in Eqs. (16.39) and (16.40). The latter are again simplified boundary conditions. The diffusion equation may also be formulated for the isotropic part, $H_\omega(r)$, of the distribution function [Eq. (14.8)]. In this case, the boundary conditions are given by Eqs. (16.36) and (16.38), respectively.

For practical calculations it is useful to dispose of a variational principle. Such a principle was formulated for the GL approximation in Section 15d, and for the diffusion approximation at the end of Section 16b.

It is not our aim to present a critical survey of the literature concerned with the problem of metals in electric contact. Only a simple limiting case shall be treated with the help of the methods devised in this review.

### b) Cooper Limit

Let us imagine a sample consisting of two superposed films of thickness $a_1$ and $a_2$, respectively. The diffusion approximation shall be valid in both films. That means that the electronic mean free path is sufficiently short so that Eq. (14.15) holds in both metals. Further, the electronic mean free path has to be short as compared to the film thickness. It is now assumed that both films are thin enough so that the isotropic part of the distribution function, $H_\omega(r)$, in the sense of Eq. (14.1), is approximately constant inside each film. This assumption shall be discussed critically below in small print. Further, it is assumed in the main text that the Debye frequencies of both metals are equal and that it suffices to work with the simplified boundary condition (16.38) on the interface. On the two external surfaces, boundary condition (16.2) is, of course, valid.

Let us first calculate the transition temperature [66] of the sample in the absence of a magnetic field. If $H_\omega(r)$ is constant inside each metal, also the pair potential is constant inside each metal as a consequence of Eq. (14.9). We denote the value by $\Delta_1$ and $\Delta_2$, respectively. For the simplified boundary conditions, Eq. (16.43) leads to the result that the (constant) isotropic part of the distribution function is the same in both metals. It shall be denoted by $H_\omega$.

The calculation is now based upon the variational principle connected with the functional (16.41), where the boundary integral has been discarded. If the integration is only extended along the surface and interface normal, $I'_\omega$ is replaced by

$$i'_\omega = 2|\omega|\,|H_\omega|^2\,(N_1 a_1 + N_2 a_2)$$

$$- \frac{1}{(2\pi)^2}\left[H_\omega^*(N_1 a_1 \Delta_1 + N_2 a_2 \Delta_2) + (N_1 a_1 \Delta_1^* + N_2 a_2 \Delta_2^*)\,H_\omega\right].$$

(23.1)

Putting the derivative with respect to $H_\omega^*$ equal to zero, we obtain

$$2|\omega| \, (N_1 a_1 + N_2 a_2) \, H_\omega = \frac{1}{(2\pi)^2} \, (N_1 a_1 \Delta_1 + N_2 a_2 \Delta_2) \, . \qquad (23.2)$$

Inserting $H_\omega$ into Eq. (14.9), we derive

$$\Delta_1 = 2\pi \, g_1 N_1 \, \frac{N_1 a_1 \Delta_1 + N_2 a_2 \Delta_2}{N_1 a_1 + N_2 a_2} \, T \sum_\omega{}' \, \frac{1}{2|\omega|} \qquad (23.3)$$

and an analogous equation with subscripts 1 and 2 interchanged. Forming a linear combination of the two equations in such a way that $N_1 a_1 \Delta_1 + N_2 a_2 \Delta_2$ appears also on the lefthand side, we arrive at

$$1 = 2\pi \, \frac{g_1 N_1^2 a_1 + g_2 N_2^2 a_2}{N_1 a_1 + N_2 a_2} \, T \sum_\omega{}' \, \frac{1}{2|\omega|} \, . \qquad (23.4)$$

This equation may be solved for the transition temperature, $T_c$, in the same way as this was done in Section 11c for a single metal. The result is

$$T_c = \frac{2\gamma}{\pi} \, \omega_D \exp\left[ -\frac{1}{(gN)_{eff}} \right] \qquad (23.5)$$

with

$$(gN)_{eff} = \frac{g_1 N_1^2 a_1 + g_2 N_2^2 a_2}{N_1 a_1 + N_2 a_2} \, . \qquad (23.6)$$

This result was obtained earlier by *de Gennes* [1], whereas *Cooper*'s original result [66] was slightly different.

The assumption of a position independent $H_\omega(r)$ deserves a critical comment. Let us start by assuming that $\Delta(r)$ is approximately constant in a particular metal. The film geometry then leads to the conclusion that $H_\omega(r)$ depends only upon the coordinate normal to surfaces and interface. The diffusion equation (14.8) may be solved in closed form with help of hyperbolic functions. Taking into account the boundary condition (16.2) on the external surfaces, we see that $H_\omega(r)$ is approximately constant provided that it is

$$a_j \ll \sqrt{\frac{v_j l_{jtr}}{6|\omega|}} \qquad (23.7)$$

even at the cut-off, for $|\omega|$ equal to $\omega_D$.

To be more realistic, we shall now assume that the Debye frequencies of the two metals are different, $\omega_{D1}$ and $\omega_{D2}$. Without loss of generality we may assume that it is

$$\omega_{D1} < \omega_{D2} \, . \qquad (23.8)$$

Then Eq. (23.2) is still valid in the interval $0 < |\omega| < \omega_{D1}$. But $\Delta_1$ has in this equation to be put equal to zero in the interval $\omega_{D1} < |\omega| < \omega_{D2}$. The sum in Eq. (23.3) is now extended over the interval $-\omega_{D1} < |\omega| < +\omega_{D1}$, whereas it is extended over the interval $-\omega_{D2} < |\omega| < +\omega_{D2}$ in the corresponding equation for $\Delta_2$. In this way, we again derive a system of linear equations for $\Delta_1$ and $\Delta_2$. Putting the determinant equal to zero, we obtain

$$\ln\left(\frac{2\gamma\omega_{D1}}{\pi T_c}\right) = \frac{1 - \dfrac{g_2 N_2^2 a_2}{N_1 a_1 + N_2 a_2}\ln\left(\dfrac{\omega_{D2}}{\omega_{D1}}\right)}{\dfrac{g_1 N_1^2 a_1 + g_2 N_2^2 a_2}{N_1 a_1 + N_2 a_2} - \dfrac{g_1 g_2 N_1^2 N_2^2 a_1 a_2}{(N_1 a_1 + N_2 a_2)^2}\ln\left(\dfrac{\omega_{D2}}{\omega_{D1}}\right)}. \tag{23.9}$$

If it is permitted to expand the denominator and to keep only the term linear in the logarithm, we derive

$$T_c = \frac{2\gamma}{\pi}\,\omega_{D\text{eff}}\exp\left(-\frac{1}{(gN)_{\text{eff}}}\right) \tag{23.10}$$

with $(gN)_{\text{eff}}$ still given by Eq. (23.6), and with

$$\ln\left(\frac{\omega_{D\text{eff}}}{\omega_{D1}}\right) = \left(\frac{g_2 N_2^2 a_2}{g_1 N_1^2 a_1 + g_2 N_2^2 a_2}\right)^2\ln\left(\frac{\omega_{D2}}{\omega_{D1}}\right). \tag{23.11}$$

It is, perhaps, remarkable that the last relation is not symmetric with respect to the subscripts 1 and 2.

The Debye frequencies of both metals shall now again be put equal, but the simplified boundary condition (16.38) shall be replaced by the correct one [Eq. (16.36)]. $H_\omega(r)$ is now only constant inside each metal but it may be different in different metals. We call these constants $H_{1\omega}$ and $H_{2\omega}$, respectively. Eq. (23.1) is replaced by

$$i'_\omega = 2|\omega|\,(N_1 a_1 |H_{1\omega}|^2 + N_2 a_2 |H_{2\omega}|^2)$$

$$- \frac{1}{(2\pi)^2}\,[N_1 a_1(H_{1\omega}^*\Delta_1 + \Delta_1^* H_{1\omega}) + N_2 a_2(H_{2\omega}^*\Delta_2 + \Delta_2^* H_{2\omega})] \tag{23.12}$$

$$+ B_{12}|H_{2\omega} - H_{1\omega}|^2.$$

The quantity $B_{12}$ characterizes the transmission properties of the interface. It is given by Eq. (16.42). Putting the derivatives of $i'_\omega$ with respect to $H_{1\omega}^*$ and to $H_{2\omega}^*$ equal to zero, we obtain

$$\Delta_1 = 2\pi g_1 N_1\,T\sum_\omega{}'\left[\frac{\Delta_1}{2|\omega|} + \frac{B_{12} N_2 a_2(\Delta_2 - \Delta_1)}{2|\omega|\,(2|\omega|\,N_1 N_2 a_1 a_2 + B_{12}(N_1 a_1 + N_2 a_2))}\right] \tag{23.13}$$

after some manipulations. The same equation holds with subscripts 1 and 2 interchanged. Eq. (23.4) is obtained only, if it is

$$B_{12}\left(\frac{1}{N_1 a_1} + \frac{1}{N_2 a_2}\right) \gg 2|\omega|. \tag{23.14}$$

This condition is required to hold even for $|\omega|$ equal to $\omega_D$.

It is seen from the considerations presented above in small print, that rather stringent conditions have to be satisfied in order that the simple result (23.5) is valid.

Now the surface normal critical field, $H_{c\perp}(T)$, of the sample shall be calculated [67]. We let the $z$ direction of a Cartesian coordinate system coincide with the surface and interface normal and use the gauge (19.1). The general considerations of Section 20b are still valid. We

14*

therefore put

$$\Delta(r) = \Delta_j f(x) \qquad (23.15)$$

with the function $f(x)$ being given by Eq. (19.13). The subscript, $j$, refers to the metal in question. In the spirit of the Cooper approximation, we further put

$$H_\omega(r) = H_\omega f(x) \qquad (23.16)$$

where Eq. (16.43) has already been taken into account. Eqs. (23.15) and (23.16) are compatible in the sense of Eq. (14.9).

In the functional (16.41), now also the derivative term has to be taken into account. The spatial integration has to be extended both in $z$ and in $x$ direction. We, therefore, put

$$i'_\omega = 2|\omega| \, |H_\omega|^2 \, (N_1 a_1 + N_2 a_2) \int\limits_{-\infty}^{+\infty} f^2(x) \, \mathrm{d}x$$

$$+ \frac{1}{3} |H_\omega|^2 \, (N_1 a_1 v_1 l_{1\,\mathrm{tr}} + N_2 a_2 v_2 l_{2\,\mathrm{tr}}) \int\limits_{-\infty}^{+\infty} \left[ \left( \frac{\mathrm{d}f(x)}{\mathrm{d}x} \right)^2 + (eHx)^2 f^2(x) \right] \mathrm{d}x \qquad (23.17)$$

$$- \frac{1}{(2\pi)^2} \left[ H_\omega^*(N_1 a_1 \Delta_1 + N_2 a_2 \Delta_2) + (N_1 a_1 \Delta_1^* + N_2 a_2 \Delta_2^*) H_\omega \right] \int\limits_{-\infty}^{+\infty} f^2(x) \, \mathrm{d}x \, .$$

Here, it is

$$\int\limits_{-\infty}^{+\infty} \left[ \left( \frac{\mathrm{d}f(x)}{\mathrm{d}x} \right)^2 + (eHx)^2 \, f^2(x) \right] \mathrm{d}x = 2eH \int\limits_{-\infty}^{+\infty} f^2(x) \, \mathrm{d}x, \qquad (23.18)$$

as follows from the differential equation for $f(x)$ [Eq. (19.6) with $k=0$, $n=0$]. Putting $\partial i'_\omega / \partial H_\omega^*$ equal to zero, we obtain

$$\left[ 2|\omega| \, (N_1 a_1 + N_2 a_2) + \frac{2eH}{3} (N_1 a_1 v_1 l_{1\,\mathrm{tr}} + N_2 a_2 v_2 l_{2\,\mathrm{tr}}) \right] H_\omega$$

$$= \frac{1}{(2\pi)^2} (N_1 a_1 \Delta_1 + N_2 a_2 \Delta_2), \qquad (23.19)$$

which replaces Eq. (23.2). Instead of Eq. (23.3), we find

$$\Delta_1 = 2\pi g_1 N_1 \frac{N_1 a_1 \Delta_1 + N_2 a_2 \Delta_2}{N_1 a_1 + N_2 a_2} \, T \sum_\omega{}' \frac{1}{2|\omega| + \dfrac{2eH}{3} (v l_{\mathrm{tr}})_{\mathrm{eff}}} \qquad (23.20)$$

with

$$(v l_{\mathrm{tr}})_{\mathrm{eff}} = \frac{N_1 a_1 v_1 l_{1\,\mathrm{tr}} + N_2 a_2 v_2 l_{2\,\mathrm{tr}}}{N_1 a_1 + N_2 a_2} \, . \qquad (23.21)$$

Eq. (23.4) is replaced by

$$1 = 2\pi \frac{g_1 N_1^2 a_1 + g_2 N_2^2 a_2}{N_1 a_1 + N_2 a_2} T \sum_{\omega}' \frac{1}{2|\omega| + \dfrac{2eH}{3} (v \, l_{tr})_{eff}} . \quad (23.22)$$

Comparing this equation with, e.g., Eq. (14.30), we see that it is

$$2 e H_{c\perp}(T) = \frac{3\mu(T)}{(v \, l_{tr})_{eff}} , \quad (23.23)$$

with $\mu(T)$ defined by Eq. (14.33) and represented graphically in Fig. 2. $T_c$ has, in these relations, to be identified with the transition temperature of the sample in the absence of a magnetic field, as given by Eq. (23.5).

The surface parallel critical field, $H_{c\parallel}(T)$, may be calculated in an analogous manner [68]. Now we let the $x$ axis coincide with the surface normal and again use the gauge (19.1). We assume metal 1 to be extended from $-a_1$ to 0, and metal 2 from 0 to $+a_2$. We further put

$$\Delta(r) = \Delta_j \exp(iky) \quad (23.24)$$

and

$$H_\omega(r) = H_\omega \exp(iky) , \quad (23.25)$$

where the parameter, $k$, remains to be determined. Eq. (23.1) is replaced by

$$i_\omega' = \left[ 2|\omega| (N_1 a_1 + N_2 a_2) + \frac{1}{3} \left( N_1 v_1 l_{1\,tr} \int_{-a_1}^{0} (k + 2eHx)^2 \, dx \right. \right.$$

$$\left. \left. + N_2 v_2 l_{2\,tr} \int_0^{+a_2} (k + 2eHx)^2 \, dx \right) \right] |H_\omega|^2 \quad (23.26)$$

$$- \frac{1}{(2\pi)^2} [H_\omega^* (N_1 a_1 \Delta_1 + N_2 a_2 \Delta_2) + (N_1 a_1 \Delta_1^* + N_2 a_2 \Delta_2^*) H_\omega] .$$

Experience gained in the previous calculations immediately shows that it is

$$3\mu(T) (N_1 a_1 + N_2 a_2)$$

$$= N_1 v_1 l_{1\,tr} \int_{-a_1}^{0} (k + 2eHx)^2 \, dx + N_2 v_2 l_{2\,tr} \int_0^{a_2} (k + 2eHx)^2 \, dx \quad (23.27)$$

with $\mu(T)$ given by Eq. (14.33). In that equation, $T_c$ is again the transition temperature of the sample as written down in Eq. (23.5). The parameter $k$ in the quadratic form on the righthand side of Eq. (23.27), has now to be chosen in such a way that the magnetic field, $H$, becomes maximal.

The result is

$$(2\,e\,H_{c\parallel}(T))^2 \tag{23.28}$$

$$= 36\,\mu(T)\,\frac{(N_1 a_1 + N_2 a_2)\,(N_1 v_1 l_{1\,\mathrm{tr}} a_1 + N_2 v_2 l_{2\,\mathrm{tr}} a_2)}{(N_1 v_1 l_{1\,\mathrm{tr}} - N_2 v_2 l_{2\,\mathrm{tr}})\,(N_1 v_1 l_{1\,\mathrm{tr}} a_1^4 - N_2 v_2 l_{2\,\mathrm{tr}} a_2^4) + N_1 v_1 l_{1\,\mathrm{tr}} N_2 v_2 l_{2\,\mathrm{tr}}(a_1 + a_2)^4}\,.$$

Combining Eqs. (23.23) and (23.28), we obtain the result that $H_{c\,\parallel}^2(T)/H_{c\perp}(T)$ is temperature independent for such very thin samples!

## References

1. *de Gennes, P. G.:* Rev. Mod. Phys. **36**, 225 (1964).
2. *Lüders, G.:* Z. Naturforsch. **21a**, 680 (1966).
3. — Z. Naturforsch. **21a**, 1425 (1966).
4. — Z. Naturforsch. **21a**, 1842 (1966).
5. — Z. Naturforsch. **22a**, 845 (1967).
6. — Z. Naturforsch. **23a**, 1 (1968).
7. — Z. Naturforsch. **21a**, 1415 (1966).
8. *Ginzburg, V. L., Landau, L. D.:* Zh. Eksperim. Teor. Fiz. (USSR) **20**, 1064 (1950); Engl. transl. in Men of physics: *L. D. Landau*, (Editor: *D. ter Haar*), London: Pergamon Press, 1965. German transl. Phys. Abh. Sowjetunion, Folge **1**, S. 1, Leipzig 1958.
9. *Maki, K.:* Physics **1**, 21 (1964).
10. *de Gennes, P. G.:* Phys. Kond. Mat. **3**, 79 (1964).
11. *Eilenberger, G.:* Z. Physik **214**, 195 (1968).
12. *Schöler, E.:* Thesis Göttingen 1966, unpublished.
13. *Usadel, K. D.:* Thesis Göttingen 1967; Z. Naturforsch. **23a**, 655 (1968).
14. — *Schmidt, M.:* Z. Physik **221**, 35 (1969).
15. *de Gennes, P. G., Tinkham, M.:* Physics **1**, 107 (1964).
16. *Shapoval, E. A.:* Zh. Eksperim. Teor. Fiz. (USSR) **47**, 1007 (1964); Engl. transl. Soviet Phys. JETP **20**, 675 (1965).
17. — Zh. Eksperim. Teor. Fiz. (USSR) **49**, 930 (1965); Engl. transl. Soviet Phys. JETP **22**, 647 (1966).
18. — Zh. Eksperim. Teor. Fiz. (USSR) **51**, 669 (1966); Engl. transl. Soviet Phys. JETP **24**, 443 (1967).
19. *Zaitsev, R. O.:* Zh. Eksperim. Teor. Fiz. (USSR) **48**, 644 (1965); Engl. transl. Soviet Phys. JETP **21**, 426 (1965).
20. — Zh. Eksperim. Teor. Fiz. (USSR) **48**, 1759 (1965); Engl. transl. Soviet Phys. JETP **21**, 1178 (1965).
21. *Thompson, R. S.:* Zh. Eksperim. Teor. Fiz. (USSR) **53**, 759 (1968); Engl. transl. Soviet Phys. JETP **26**, 470 (1968).
22. *Bardeen, J., Cooper, L. N., Schrieffer, J. R.:* Phys. Rev. **108**, 1175 (1957).
23. *Pippard, A. B.:* Proc. Roy. Soc. A **216**, 547 (1953).
24. *Gorkov, L. P.:* Zh. Eksperim. Teor. Fiz. (USSR) **34**, 735 (1959); Engl. transl. Soviet Phys. JETP **9**, 636 (1959).
25. *Bogoliubov, N. N.:* Physica **26**, S 1 (1960).
26. *Eilenberger, G., Ambegaokar, V.:* Phys. Rev. **158**, 332 (1967).
27. *Reuter, G. E. H., Sondheimer, E. H.:* Proc. Roy. Soc. A **195**, 336 (1948).
28. *Usadel, K. D.:* Diplomarbeit Göttingen 1966, unpublished.
29. *Edwards, S. F.:* Phil. Mag. **3**, 1020 (1958).

30. *Abrikosov, A. A., Gorkov, L. P.:* Zh. Eksperim. Teor. Fiz. (USSR) **35**, 1558 (1958); Engl. transl. Soviet Phys. JETP **8**, 1090 (1959).
31. — Zh. Eksperim. Teor. Fiz. (USSR) **47**, 720 (1964); Engl. transl. Soviet Phys. JETP **20**, 480 (1965).
32. *Hu, C. R., Korenman, V.:* Phys. Rev. **178**, 684 (1969).
33. *Ziman, J. M.:* Principles of the theory of solids. Ch. 6. Cambridge: University Press 1965.
34. *Peierls, R.:* Z. Physik **80**, 763 (1933).
35. *Luttinger, J. M.:* Phys. Rev. **84**, 814 (1951).
36. *Abrikosov, A. A., Gorkov, L. P.:* Zh. Eksperim. Teor. Fiz. (USSR) **39**, 1781 (1960); Engl. transl. Soviet Phys. JETP **12**, 1243 (1961).
37. *Kondo, J.:* Progr. Theor. Phys. **32**, 37 (1964).
38. *Zittartz, J., Müller-Hartmann, E.:* Z. Physik **232**, 11 (1970).
39. *Sommers, H. J.:* Diplomarbeit Göttingen 1968, unpublished.
40. *Gorkov, L. P.:* Zh. Eksperim. Teor. Fiz. (USSR) **36**, 1918 (1959); Engl. transl. Soviet Phys. JETP **9**, 1364 (1959).
41. — Zh. Eksperim. Teor. Fiz. (USSR) **37**, 1407 (1959); Engl. transl. Soviet Phys. JETP **10**, 998 (1960).
42. *Harms, K. D.:* Thesis Göttingen 1970; Z. Naturforsch. **25a**, 1161 (1970).
43. *Tewordt, L.:* Phys. Rev. **137**, A 1745 (1965).
44. *Steppeler, J.:* Diplomarbeit Göttingen 1968, unpublished.
45. *Schmidt, M.:* Diplomarbeit Göttingen 1969, unpublished.
46. *Rickayzen, G.:* Phys. Rev. **138**, A 73 (1965).
47. *Abrikosov, A. A.:* Zh. Eksperim. Teor. Fiz. (USSR) **32**, 1442 (1957); Engl. transl. Soviet Phys. JETP **5**, 1174 (1957).
48. *Helfand, E., Werthamer, N. R.:* Phys. Rev. Letters **13**, 686 (1964); Phys. Rev. **147**, 288 (1966).
49. *Hohenberg, P. C., Werthamer, N. R.:* Phys. Rev. **153**, 493 (1967).
50. *Gruenberg, L. W., Gunther, L.:* Phys. Rev. **176**, 606 (1968).
51. *Eilenberger, G.:* Z. Physik **190**, 142 (1966).
52. *Richter, P.:* Diplomarbeit Göttingen 1968, unpublished.
53. *Sommers, H. J.:* Thesis Göttingen 1971; to be published.
54. *Saint-James, D., de Gennes, P. G.:* Phys. Letters **7**, 306 (1963).
55. *Lüders, G.:* Z. Physik **202**, 8 (1967).
56. *Ebneth, G., Tewordt, L.:* Z. Physik **185**, 421 (1965).
57. *Lüders, G., Usadel, K. D.:* Z. Physik **222**, 358 (1969).
58. *Sarma, P. K.:* Thesis Köln 1969; Z. Physik, to be published.
59. *Hu, C. R., Korenman, V.:* Phys. Rev. **185**, 672 (1969).
60. — Thesis University of Maryland 1968, unpublished.
61. *Schultens, H. A.:* Z. Physik **232**, 430 (1970).
62. *Lüders, G.:* Z. Physik **209**, 219 (1968); Erratum: Z. Physik **214**, 108 (1968).
63. *Usadel, K. D.:* Z. Physik **215**, 77 (1968).
64. *Thompson, R. S., Baratoff, A.:* Phys. Rev. **167**, 361 (1968).
65. *Harms, K. D.:* Diplomarbeit Göttingen 1968, unpublished.
66. *Cooper, L. N.:* Phys. Rev. Letters **6**, 689 (1961).
67. *Helms, A.:* Diplomarbeit Göttingen 1970, unpublished.
68. *Usadel, K. D.:* Z. Physik **213**, 70 (1968).

Prof. Dr. *Gerhart Lüders* and Dr. *Klaus-Dieter Usadel*
Institut für Theoretische Physik der Universität
D-3400 Göttingen
Germany